POST-DIGITAL ELECTRONICS

THE ELLIS HORWOOD SERIES IN COMPUTERS AND THEIR APPLICATIONS

Series Editor: BRIAN MEEK

Director of the Computer Unit, Queen Elizabeth College, University of London

The series aims to provide up-to-date and readable texts on the theory and practice of computing, with particular though not exclusive emphasis on computer applications. Preference is given in planning the series to new or developing areas, or to new approaches in established areas.

The books will usually be at the level of introductory or advanced undergraduate courses. In most cases they will be suitable as course texts, with their use in industrial and commercial fields always kept in mind. Together they will provide a valuable nucleus for a computing science library.

INTERACTIVE COMPUTER GRAPHICS IN SCIENCE TEACHING
Edited by J. McKENZIE, University College, London, L. ELTON, University of Surrey, R. LEWIS, Chelsea College, London.
INTRODUCTORY ALGOL 68 PROGRAMMING
D. F. BRAILSFORD and A. N. WALKER, University of Nottingham.
GUIDE TO GOOD PROGRAMMING PRACTICE
Edited by B. L. MEEK, Queen Elizabeth College, London and P. HEATH, Plymouth Polytechnic.
CLUSTER ANALYSIS ALGORITHMS: For Data Reduction and Classification of Objects
H. SPÄTH, Professor of Mathematics, Oldenburg University.
DYNAMIC REGRESSION: Theory and Algorithms
L. J. SLATER, Department of Applied Engineering, Cambridge University and H. M. PESARAN, Trinity College, Cambridge
FOUNDATIONS OF PROGRAMMING WITH PASCAL
LAWRIE MOORE, Birkbeck College, London.
PROGRAMMING LANGUAGE STANDARDISATION
Edited by B. L. MEEK, Queen Elizabeth College, London and I. D. HILL, Clinical Research Centre, Harrow.
THE DARTMOUTH TIME SHARING SYSTEM
G. M. BULL, The Hatfield Polytechnic
RECURSIVE FUNCTIONS IN COMPUTER SCIENCE
R. PETER, formerly Eötvos Lorand University of Budapest.
FUNDAMENTALS OF COMPUTER LOGIC
D. HUTCHISON, University of Strathclyde.
THE MICROCHIP AS AN APPROPRIATE TECHNOLOGY
Dr. A. BURNS, The Computing Laboratory, Bradford University
SYSTEMS ANALYSIS AND DESIGN FOR COMPUTER APPLICATION
D. MILLINGTON, University of Strathclyde.
COMPUTING USING BASIC: An Interactive Approach
TONIA COPE, Oxford University Computing Teaching Centre.
RECURSIVE DESCENT COMPILING
A. J. T. DAVIE and R. MORRISON, University of St. Andrews, Scotland.
PASCAL IMPLEMENTATION
S. PEMBERTON and M. DANIELS, Brighton Polytechnic
MICROCOMPUTERS IN EDUCATION
Edited by I. C. H. SMITH, Queen Elizabeth College, University of London
AN INTRODUCTION TO PROGRAMMING LANGUAGE TRANSITION
R. E. BERRY, University of Lancaster
ADA: A PROGRAMMER'S CONVERSION COURSE
M. J. STRATFORD-COLLINS, U.S.A.
STRUCTURED PROGRAMMING WITH COMAL
R. ATHERTON, Bulmershe College of Higher Education
SOFTWARE ENGINEERING
K. GEWALD, G. HAAKE and W. PFADLER, Siemens AG, Munich

17-95
ML

This book is to be returned on or before
... date stamped below.

STRATHCLYDE UNIVERSITY LIBRARY
30125 00064009 3

POST-DIGITAL ELECTRONICS

F. R. PETTIT, C.Eng.
Head of Computing Teaching Centre
University of Oxford

ELLIS HORWOOD LIMITED
Publishers · Chichester

Halsted Press: a division of
JOHN WILEY & SONS
New York · Brisbane · Chichester · Toronto

UNIVERSITY OF
STRATHCLYDE LIBRARIES

05696397

First published in 1982 by
ELLIS HORWOOD LIMITED
Market Cross House, Cooper Street, Chichester, West Sussex, PO19 1EB, England

The publisher's colophon is reproduced from James Gillison's drawing of the ancient Market Cross, Chichester.

Distributors:

Australia, New Zealand, South-east Asia:
Jacaranda-Wiley Ltd., Jacaranda Press,
JOHN WILEY & SONS INC.,
G.P.O. Box 859, Brisbane, Queensland 40001, Australia

Canada:
JOHN WILEY & SONS CANADA LIMITED
22 Worcester Road, Rexdale, Ontario, Canada.

Europe, Africa:
JOHN WILEY & SONS LIMITED
Baffins Lane, Chichester, West Sussex, England.

North and South America and the rest of the world:
Halsted Press: a division of
JOHN WILEY & SONS
605 Third Avenue, New York, N.Y. 10016, U.S.A.

©1982 F. R. Pettit/Ellis Horwood Ltd., Publishers

British Library Cataloguing in Publication Data
Pettit, F. R.
Post digital electronics
1. Pulse circuits
I. Title
621.3815'34 TK7868.P8

ISBN 0-85312-421-3 (Ellis Horwood Ltd., Publishers)
ISBN 0-470-27334-8 (Halsted Press)

Typeset in Press Roman by Ellis Horwood Ltd.
Printed in Great Britain by R. J. Acford, Chichester.

COPYRIGHT NOTICE –
All Rights Reserved. No part of this publication may be reproduced, stored in a retrieval system, or transmitted, in any form or by any means, electronic, mechanical, photocopying, recording or otherwise, without the permission of Ellis Horwood Limited, Market Cross House, Cooper Street, Chichester, West Sussex, England.

D
621.3815'34
PET

Table of Contents

Foreword 11

Chapter 1 Reliability Concepts and the Positive Principle

1.1 Reliability, Availability and the Utility Concept 13
1.2 Improvements in Performance During 1930–1980 23
1.3 Improvements in Components – the Concept of the Component . . 25
1.4 Improvements in Interconnections 25
1.5 Reduction in Interconnections – the LSI Microelectronic Device 26
1.6 Failure Modes in Components 27
1.6.1 Degradation 27
1.6.2 Solid Failure 27
1.6.3 Intermittent Failure 27
1.7 Effects of Component Failures 28
1.7.1 Effects of Failures in Analog Equipment 28
1.7.2 Effects of Failures in Digital Equipment 28
1.8 Techniques of System Operation 29
1.9 Use of Redundancy in System Design 30
1.10 Use of Auto-test Circuitry 30
1.11 Response of Systems to Noisy Signals 32
1.12 Processes and Processors 33
1.13 Distributed-function Processors 33
1.14 Limits to Complexity – Availability 34
1.15 Reliability Variation During Operational Life 34
1.16 Numbers Information, Measure and Measurement 35
1.17 The Positive Principle in Reliability 36

Chapter 2 Geometrical Pulsator Matrices Fundamental Theory

2.1 The search commences 37

2.2 The AND–END element 37
2.3 Pulsorcubes 37
2.4 Pulsorcube interconnections 38
2.5 Analysis of the synchronous pulsorcube 42
2.6 Analysis of the asynchronous pulsorcube 43
2.7 The superpulsorcube 45
2.8 The pulsorhypercube 45
2.9 Logic function versus activity 48
2.10 Multidimensional electronic circuitry 49
2.11 Pulsor sensors 50
2.12 A multiplanar pulsator sensor-activator 50
2.13 The positive principal in action 54

Chapter 3 From Logic to Probability

3.1 Approaches to Analysis 55
3.2 Logic-to-Probability Transformation 55
3.2.1 Logical Symbolism 55
3.2.2 Logical Constants 56
3.2.3 Logical Variables 56
3.2.4 Logical Operators 56
3.2.5 Pulse Rate, Frequency and Probability 57
3.3 Combinatorial Logic and Probability Transforms 57
3.4 Further Notes on Logical Analysis – De Morgan's Theorem 59
3.5 Probabilistic Transfer Functions 59
3.6 Probabilistic Equations 59
3.7 Probabilistic Analyses 60
3.7.1 The Synchronous Two-input AND 60
3.7.2 The Synchronous Two-input OR 61
3.7.3 The Asynchronous Two-input AND–END 61
3.7.4 The Asynchronous Two-input OR–END 61
3.7.5 The Asynchronous Three-input Majority-logic Element 61
3.8 Conclusions Regarding Logic Processor Elements 62
3.9 The Leaky Summator Element 62
3.9.1 The Summator – Probability Gain 62
3.9.2 Leaky Summators 63
3.9.3 The Leaky Summator Element – Probability Gain 63
3.10 Application Example – Regular four-input LSEs 64
3.10.1 A Right Pyramidal Matrix of LSEs 64
3.10.2 The Effect of a Bright Spot at Sensor Centre 66
3.10.3 An Off-centre Spot 67
3.10.4 A Defocussed Spot 67
3.11 Characteristics of the LSE 69

Chapter 4 Elemental Pulsator Devices

4.1 Footnote to all circuits in this Chapter 70
4.2 Some elementary Logic Circuits 70
4.3 An RTL Circuit 71
4.4 Pulse Delay Elements 73
4.5 Leaky Summator Circuits 75
4.5.1 Quiescent State 76
4.6 LSE Fabrication 77
4.7 Inhibition 78
4.8 LSE with Extended Features 78
4.9 LSEs as Sensors 80
4.10 Electrochemical Cells 80
4.11 Test Pattern Generators 80

Chapter 5 Systematic Propagative Matrices

5.1 Geometrical Matrices 82
5.2 Logical Cubic Lattices 83
5.3 The Point-source Cubic 84
5.4 Plane-source Cubic (Quad-input Elements) 86
5.5 Divergence, Convergence and Bivergence 87
5.6 Triple-input Plane-source Tetrahedral matrix 87
5.7 Triple-output Point-source Tetrahedron 88
5.8 Quad-input tetrahedron with plane bypass 89
5.9 Plane-source Tubular Matrices 89

Chapter 6 Application of the Positive Principal

6.1 Towards Reliability – Desystemetization 93
6.2 Definition of terms 94
6.2.1 Plane Bypass (or Penetration) 94
6.2.2 Backcoupling 94
6.2.3 Reversed Elements 94
6.2.4 Intraplane Coupling 94
6.3 Design considerations 94
6.4 Interconnection Specification 95
6.5 Plane Bypass Connections 96
6.6 Penetration and Divergence (Lateral Spread) 99
6.7 Backcoupling 99
6.8 Intraplane Interconnections 101
6.9 Reverse-connected Elements 101
6.10 The Network Transfer Function 101
6.11 Inhibitory Functions 103

Chapter 7 Dynamic Processing Matrices
7.1 Processing by LSE Pulsatory Matrices 105
7.2 The Information Cloud. 105
7.3 Propagation Velocity 106
7.4 Detection of Moving Stimuli 107
7.5 Time-space Differentiation 107
7.6 The Role of Inhibition in Autostabilization 108
7.7 Shuttering 109
7.8 Selectivity and Discrimination 110
7.9 Resolution Upgrade by Jitter. 110
7.10 Dynamic Parameter Stabilization 110
7.11 Response to Noisy Signals. 111
7.12 Enhancement of Localized Feedback (Backcoupling) 111
7.13 The Effect of Defective Components. 112

Chapter 8 Linear Reflexive Memory
8.1 Graphic representation of matrix interconnections. 113
8.2 Reversed elements 113
8.2.1 Effects of Reversed Elements. 116
8.3 Factors affecting Matrix Propagation. 116
8.4 Proportion of Element Reversals 119
8.5 Amplifiers 119
8.5.1 Gain Factor 119
8.5.2 Bandwidth 121
8.5.3 Signal Enhancement by Jitter 121
8.6 The Processes of Flushing 122
8.7 Propagation Velocity in a Pulsor Matrix. 122
8.8 Velocity Modification by Elemental Reversal 123
8.8.1 Initial Velocity 123
8.8.2 Velocity in a Seeded Matrix. 123
8.8.3 Discrimination by a Seeded Matrix 123
8.9 The Interplay of Parameters 123
8.10 Interactions with External Fields 124
8.10.1 Fields Produced by Pulsor Matrices. 124
8.10.2 Sensitivity to External Fields. 125
8.10.3 Effects of Radiation and Particles. 125
8.11 Light Guides in Pulsor Processors 125
8.12 The Power of Computers 126

Chapter 9 Transverse Memory
9.1 The Transverse Memory Technique. 128
9.2 Memory Capacity 129

9.3 The Cross-section of a Circulatory Memory 130
9.4 Memory Coupling Techniques 131
9.4.1 A Transverse Injector 131
9.4.2 Orthogonal Transverse Coupling 132
9.5 Setup and Propagation of a Memory Stream 132
9.6 Tests on Circulatory Memories 133

Chapter 10 Re-Entrancy
10.1 Types of Re-entrant Matrix 134
10.2 Ladder Networks 134
10.2.1 Bi-directional Ladders 135
10.2.2 Ladders with Transverse Memories 136
10.3 Reflective Networks 137

Chapter 11 Pattern Recognition
11.1 The Processes of Correlation 139
11.2 Cross Correlation and Autocorrelation 139
11.3 Pattern Recognition Processes 140
11.4 Image Processing 140
11.5 Multilevel Memory Systems in Correlation 141
11.6 The Pulsatory Correlator 141

Chapter 12 Design Techniques – Computer Models
12.1 The Need for a Simulator 144
12.2 The Meaning of Generate in Relation to a Computer 144
12.3 Simulator = Model + Data 145
12.4 Sequencing by Computer 145
12.5 Program Control Structure 146
12.6 The State-space Matrix 146
12.7 Models of Matrix Parameters 147
12.7.1 Size 147
12.7.2 Bypass Length (penetration) 147
12.7.3 Transverse Spread 147
12.7.4 Backcoupling 148
12.7.5 Complexity 148
12.7.6 Reversed Elements 148
12.7.7 Defective Elements 148
12.8 Models of LSEs 148
12.8.1 Characteristics 148
12.8.2 Models of Sensor Elements 149
12.8.3 Models of Outputs from Pulsor Matrices 149

12.9 Graphical Presentations by the Simulator 149
12.10 PULSENET Computer Program 150
12.11 Exercizing the Simulator 150
12.12 Use of PULSENET 156
12.12.1 Checking Theories 156
12.12.2 Deducing Design Parameters 156
12.13 Further Development of PULSENET 156

Chapter 13 Fabrication Techniques
13.1 Constructional Overview 158
13.2 Silicon Structures 158
13.2.1 Interconnection Fields 160
13.2.2 Manufacturing Yield 160
13.3 Static Chemical Interconnection Fields 161
13.4 Dynamical Chemical Interconnection Fields 161
13.5 Slice Organization 162
13.5.1 Vectored Radial Propagation 162
13.5.2 Orthogonal Propagation 162
13.5.3 Spiral Propagation 164
13.5.4 Multiple Spiral Propagation 164
13.6 Electrochemical LSEs 165

Chapter 14 FORWARD
14.1 Fibroptic Image Transformations 166
14.2 Serial Data Handlers 168
14.3 Pulsor Control Systems 169
14.4 Automata – Industrial Robotics 169
14.5 The Learning Automaton 170
14.6 Pulsors in Communications 170
14.7 Self-repair? 171
14.8 Intelligent Controllers? 171

Index 173

Foreword

The concepts presented in this book had their origins in the applications of computers to certain airspace pattern recognition problems during the early 1960s. It was realized that computers many thousands of times as powerful than then existed would be needed. The immense increase in power and availability of computers since the 1960s now permits us to tackle many pattern recognition tasks. However, the computer requires us to treat such tasks as problems of extreme complexity – and we are left with systems of low reliability which present great difficulty in setting-up and fault-tracing.

Squibby (the tortoiseshell cat) was snoozing on the lawn. That wretched man disturbed a bird. Bird flew low over Squibby who quite naturally roused, leapt skywards and captured the bird in flight. That was the start of it all. How did cat perform that set of complex manoeuvres – knowing nothing at all about vector methods, differential geometry, or even arithmetic? One of the key features of the computer is that it uses representations for numbers in its attempts at information processing. Another is that it operates serially. Even the so-called 'concurrent processors' also operate in a modified serial manner.

Could we construct a processor that does not use numbers, which operates using untold thousands of associative parallel operations and which is independent of the fault conditions which naturally develop in electronic equipment? During the search for such a device, we shall pause to examine some computer programming techniques which permit the computer to model systems whose complexity greatly exceeds the storage capacity of the computer.

ACKNOWLEDGEMENTS

The author wishes to record a debt of gratitude to:

Vic Day (of Worcestershire) and Sam Weller (of Cornwall) for many enjoyable discussions on pulsors during the past 20 years. Tonia Cope (of Oxford) for encouragement during the manuscript preparation. Steve Thomas (of Oxford) for doing quite magic things with the unconventional spelling techniques used on the original manuscript. Charles Beesley (of Oxford) for all the artwork. Snowball (the white cat with black tail) for comforting a frustrated simulator designer.

1

Reliability concepts – the positive principle

1.1 RELIABILITY, AVAILABILITY AND THE UTILITY CONCEPT

Modern electronic systems include both analog and digital processes. Digital processes include what has come to be called 'random logic' (more properly 'special-purpose logic processors') and 'sequential logic', most commonly in the form of digital computers. Sadly, all such devices are subject to electrical failure in operation. The design of a complex system includes the design of validation and diagnostic facilities and procedures. To state the obvious, our devices and systems should exhibit a high degree of cost-effective reliability.

When a defect does occur, we need to restore operational capability as quickly as possible. The Mean Time Between Failures (MTBF) must be as high as possible and the Mean Time To Restore (MTTR) must be as low as possible. To generalize from the operational point of view, we require high 'availability' of the system functions. 'Availability' can be expressed as a percentage in time.

The 'usefulness' of an equipment is commonly related to its 'complexity'. Complexity, in its turn can be related to the number of components used, feedback (in its many forms) and to ergonomic design.

Consider the simple radio transmitter circa 1930 (Fig. 1.1) which was powered from an alternator to provide Interrupted Modulated Carrier Wave (IMCW) Morse transmission. Operation was simple – run the alternator and use the Morse key. Adjustment was fun: use a good receiver to set a known frequency, attempt to adjust the 'tank circuit' with the key depressed, and simultaneously attempt to adjust the 'grid coil' until maximum power was obtained. However, the device would often fail to oscillate during tuning operations – so hold a neon lamp in the hand whilst twiddling the controls – the lamp would glow when oscillation commenced. Oh yes, then use the neon lamp again for final trimming for maximum power and swing the receiver tuning to ensure correct frequency. That was not the end: next traverse the transmission lines with the neon lamp searching for 'standing waves' – if found, adjust the antenna 'stubs'

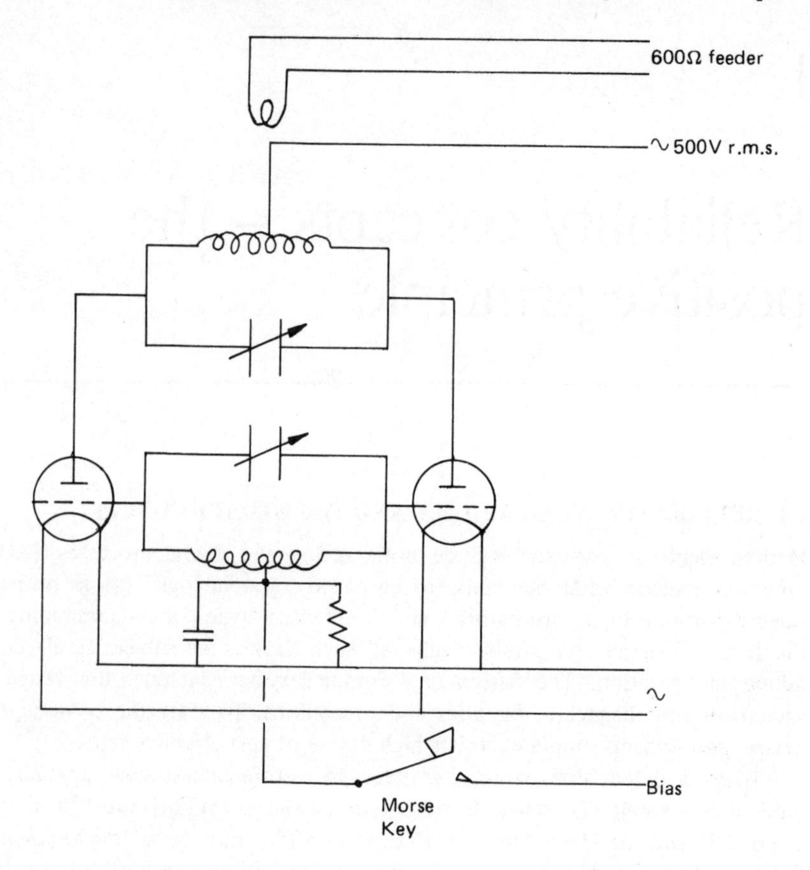

Fig. 1.1 – Radio transmitter circa 1930 – circuit diagram.

and recheck the transmitter for correct frequency and maximum power. That was the routine daily procedure for a communications station in India even into the early 1940s.

Modern transmitter designs are far more complex in component count, and their 'usefulness' is correspondingly increased. In a modern transmitter station the change of frequency can be a single operation.

The radio receiver circuit of Fig. 1.2 is almost unbelievable today but it was commonplace in 1930. One could adjust the aerial tapping on the tuning coil, the coupling between the tuning coil and the 'reaction' coil, the tuning capacitor and the magnet knob on the loudspeaker. Variations in receivers included the use of variable resistors in the valve 'filament' supply leads, additional valves, and additional 'wavetrap' coils in the aerial line. There were 'bright-emitter' and 'dull-emitter' valves and positively quaint arrangements of the coils to provide good tuning and selectivity.

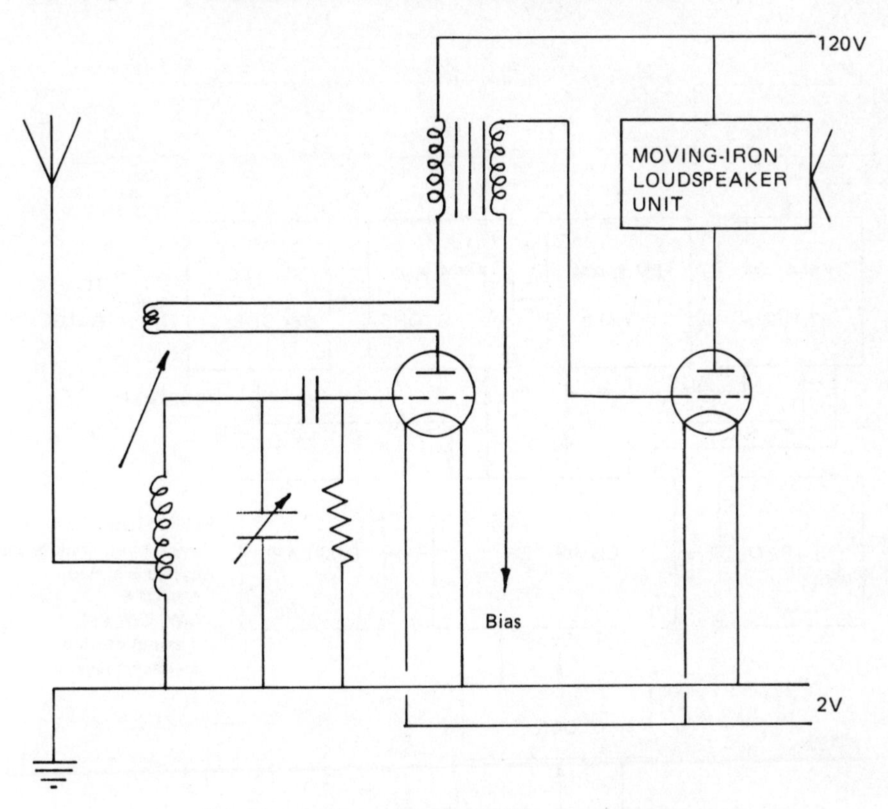

Fig. 1.2 – Radio receiver circa 1930 – circuit diagram.

Superheterodyne circuits and moving coil loudspeakers gave impressive improvements in performance and handling (usefulness) with a great increase in the number of components. But by the end of the '30s, we saw push-button automatic tuning with interstation muting and AFC, AVC, push-pull output and miltiple-loudspeaker receivers. The 'professional' market had double superhets, logging dials, variable selectivity and facilities for special communications media such as Morse – and the elements of radio navigation systems.

A simplified outline of a modern domestic receiver is given in Fig. 1.3. With the passage of time, we can expect to see the computer intrude more deeply into the signal-processing field. The audio amplifier circuits are already too complex as analog systems. The detector yields a virtually constant-amplitude signal, ideal for digitizing and the computer can replace the complex audio circuitry as a single integrated circuit component.

In modern radar systems, the numerous 'functions' of the system are dealt with by individual 'units' or 'modules'. Complexity is achieved by component count. Adjustment and servicing are relatively simple procedures. However, the

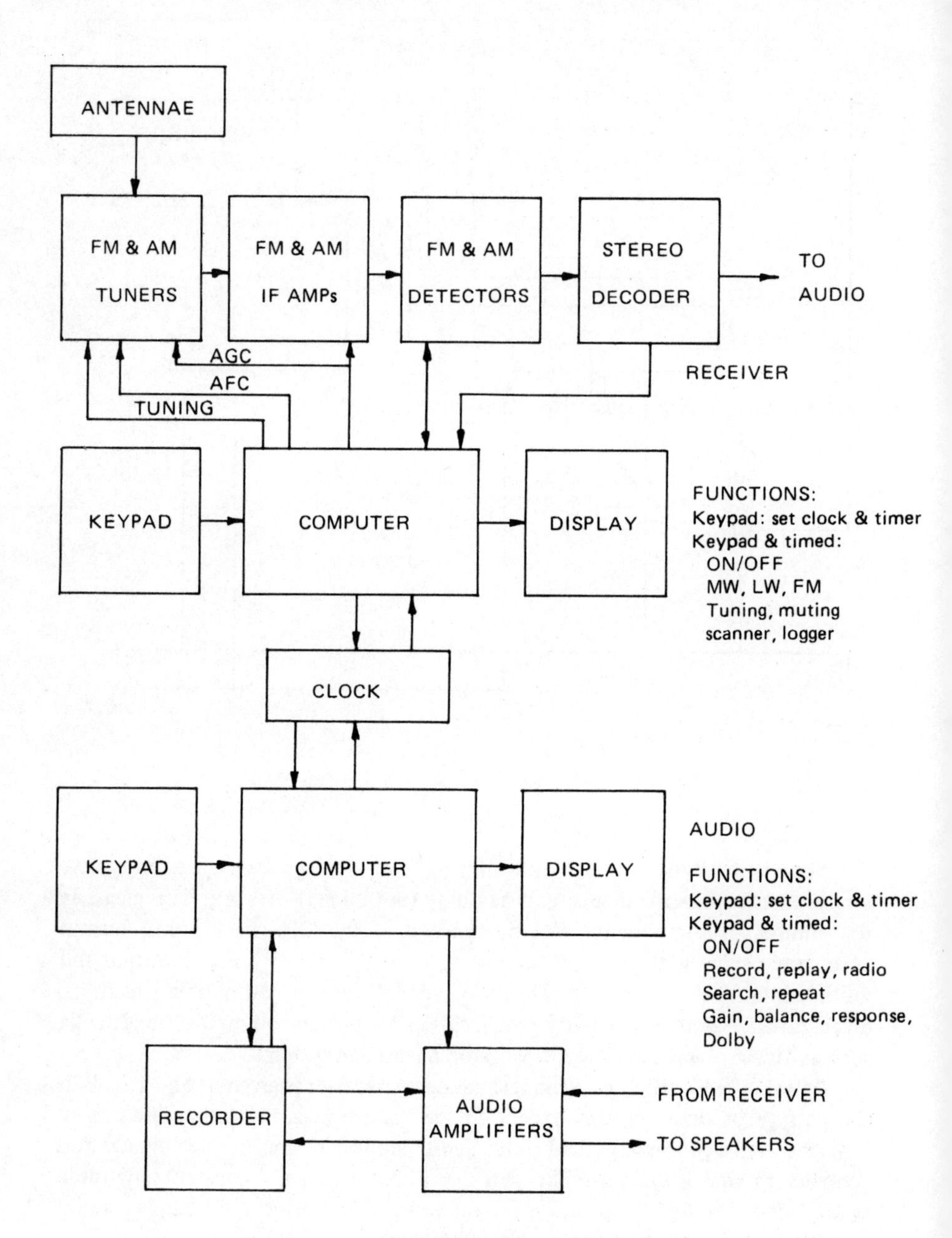

Fig. 1.3 – Scanning receiver circa 1980 – outline.

modern domestic TV receiver has a relatively low component count but houses an exceptionally complex system. Many forms of feedback are used, even to the re-use of waste power from the line-scanning circuitry.

An incredibly simple-looking circuit performs many functions such as the provision of some 20 kV EHT for the CRT and of about one third of the power needed for its own supply. The forms of complexity used in the common TV ensure very low power consumption for the whole unit, about one tenth of the consumption of the original receivers where complexity was achieved by component count.

Outline block diagrams of two radar/sonar systems are given in Figs. 1.4 and 1.5, the first showing a circa 1940 radar, the other outlining a sonar design by the author of circa 1980. The height-layered design of 1980 includes image analysis and full graphical display of those items of particular interest to the users.

This represents a design of considerable usefulness achieved by employment of a large number of single-chip computers. Complexity in terms of component count is of the order of 10^7.

The concept of increased 'usefulness' with increased complexity can again be seen in radio direction finding equipments over the 50-year period of this brief review.

The loop/sense antenna device (Fig. 1.6) of the early '30s was designed for manual operation: first the wanted transmission was selected using the sense

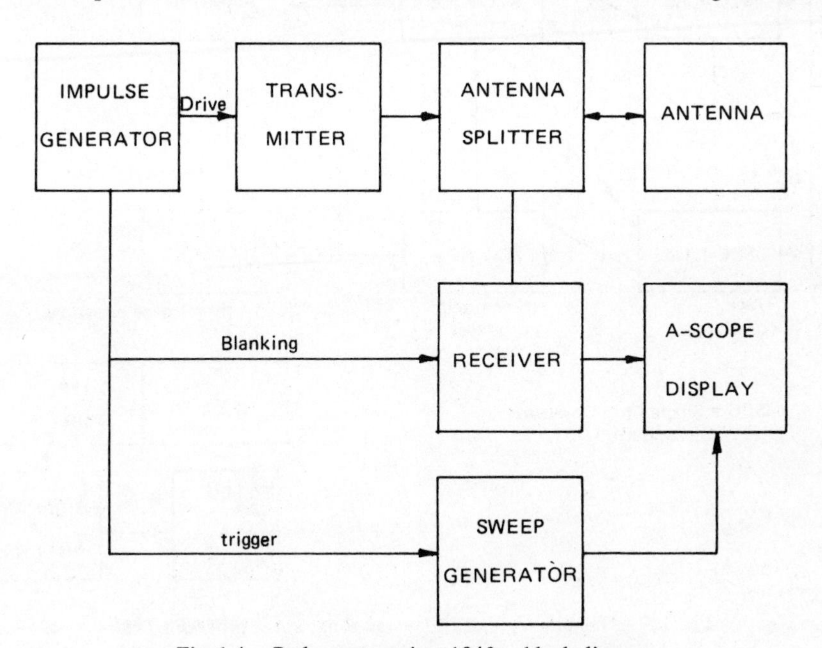

Fig. 1.4 – Radar system circa 1940 – block diagram.

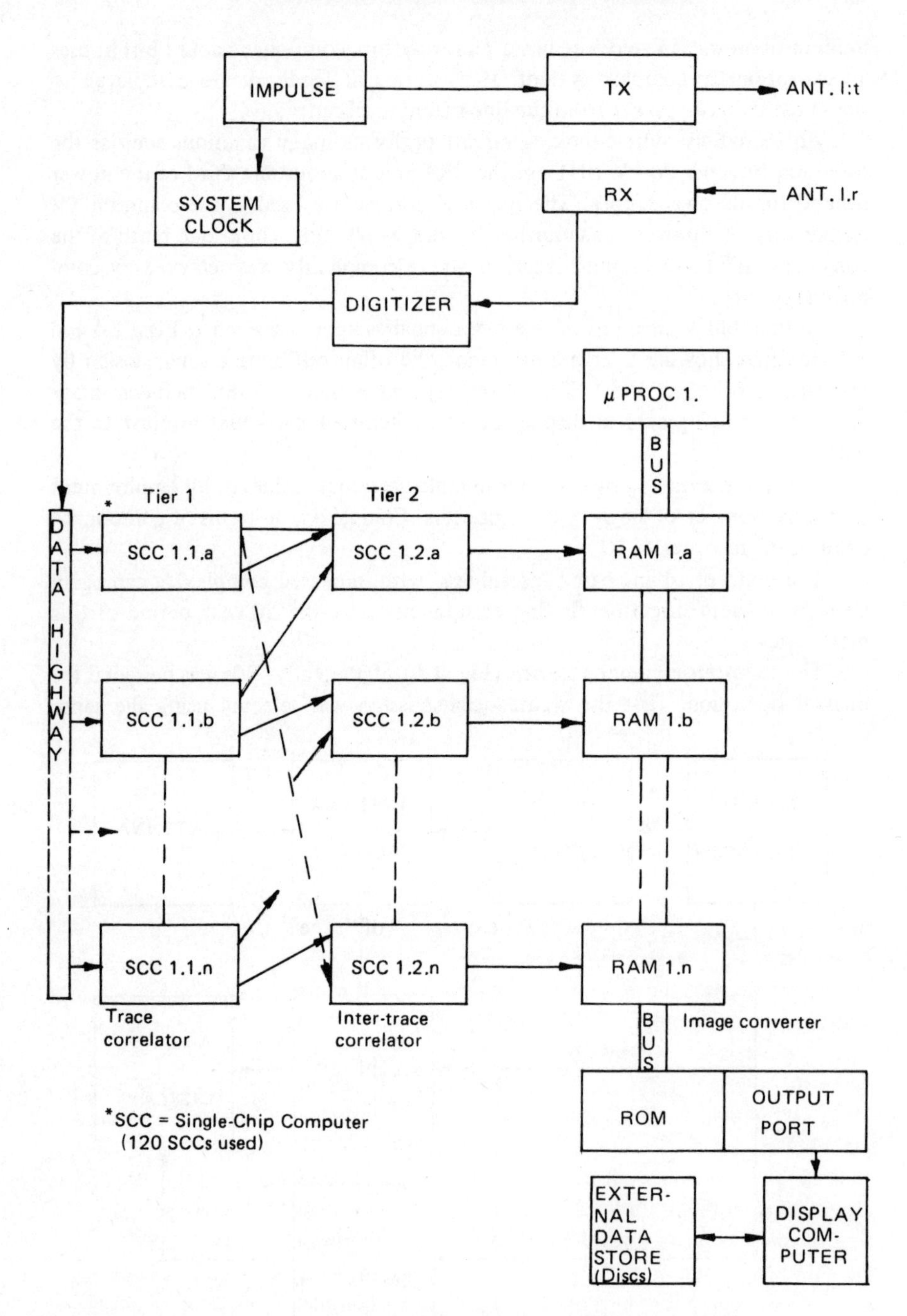

Fig. 1.5 – One plane of height-layered sonar image processor 1980.

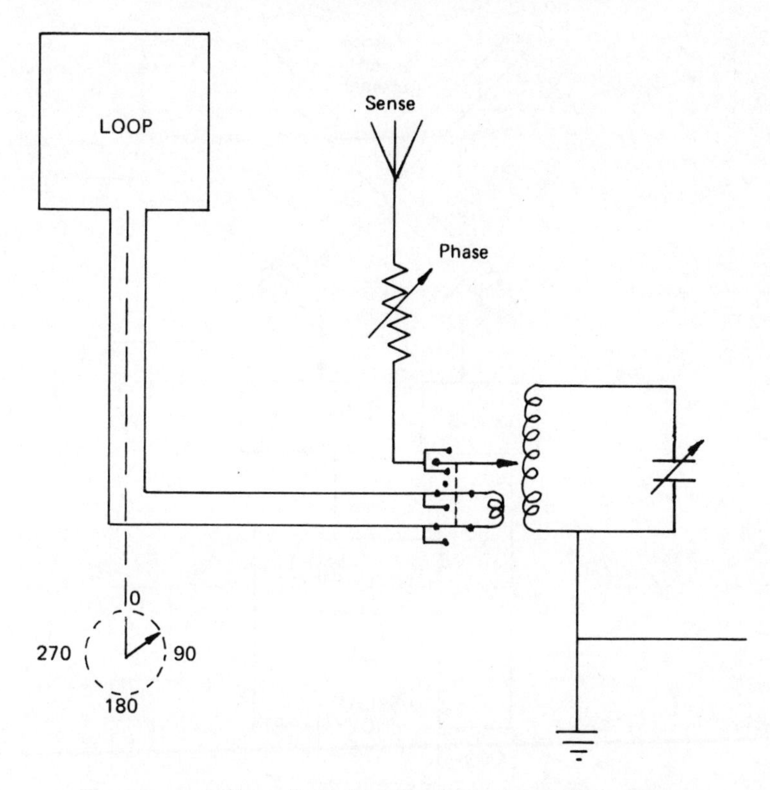

Fig. 1.6 – Loop/Sense Radio Directionfinder circa 1930 – circuit.

aerial; switch to the loop and swing to find the directions of the minimum signal strength positions (normally 180° apart), then switch-in the sense aerial, swing the loop ± 90° to determine the direction of the new 'cardioidal' minimum. After a relatively straightforward piece of juggling with the antenna switch, the loop rotation and sense controls would enable the direction of the incoming signal to be determined – unless the signal was ICW Morse transmissions with extended breaks.

The contemporary radio-compass form of Automatic Direction Finder (ADF) combined with reliable ground beacons provides a highly ergonomized navaid. Fig. 1.7 shows the ADF to be very complex in component count. The modular sub-assemblies ensure rapid fault-tracing in the event of a failure.

It is generally found that to increase the usefulness of the equipment, its complexity has to be increased. The more we require of a system, the greater the number of functions that have to be built into the system. The computer alters this in some measure by permitting more complex use of the same hardware – the complexity can often be moved from hardware to software with more or less dramatic results.

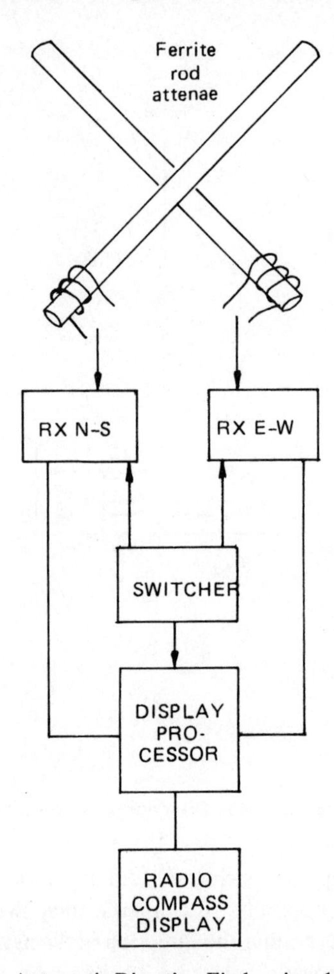

Fig. 1.7 – Automatic Direction Finder circa 1980 – outline.

Combining the concepts of availability with usefulness of a system we can assess the utility provided by the system. However, to increase the utility, complexity must generally increase, leading all too frequently to greater unreliability and hence reduced availability for operational use. We shall pursue these ideas in greater detail, examining the most popular techniques used to achieve high values of utility in functional systems.

One particularly interesting example of variation of MTBF due to maintenance procedures in a very large system is the old NASA APOLLO Remote Site Data Processor (RSPD). An outline block diagram (Fig. 1.8) indicates the number of major equipments used. A total count of some 30,000 PCBs was found in the installation plus many electro-mechanical devices, such as printers, plotters, magtape drives, papertape handlers, and flight controllers' consoles.

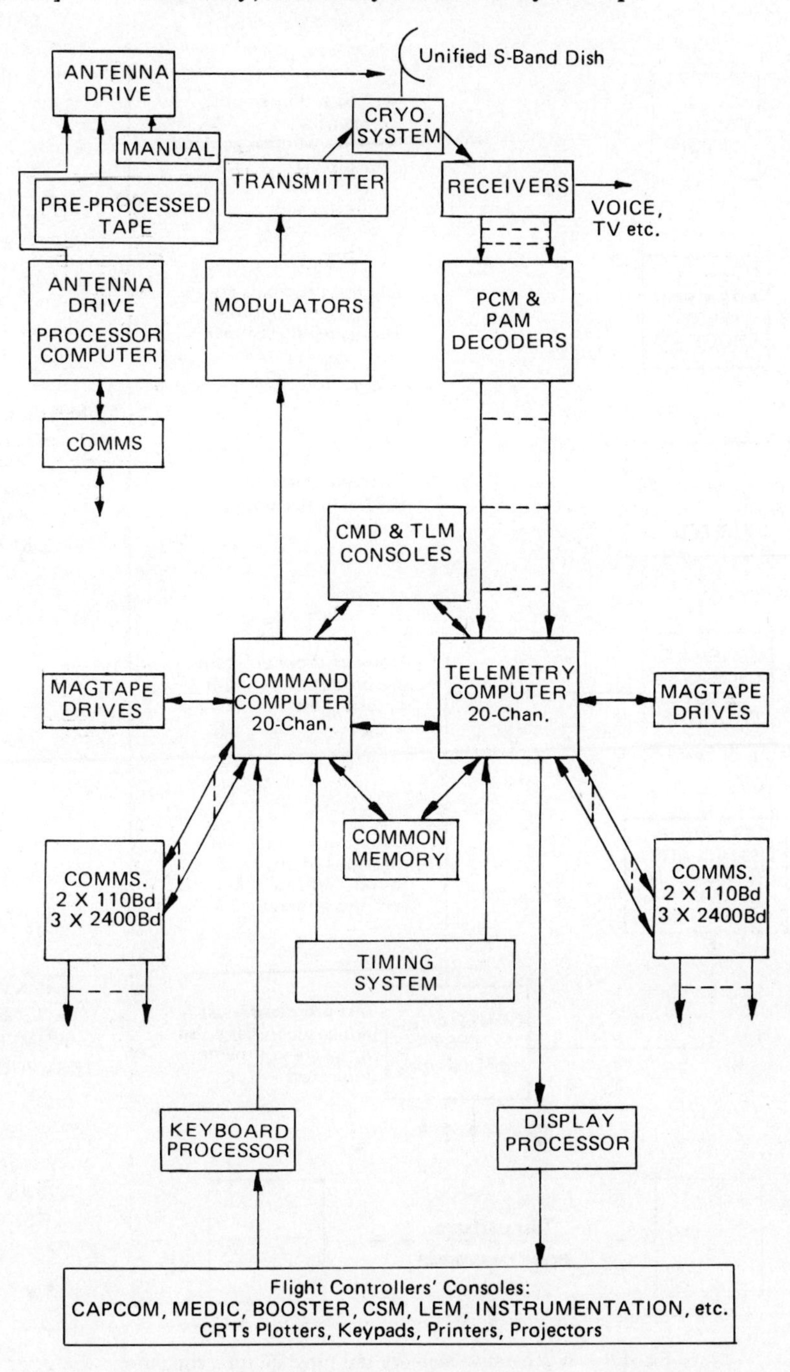

Fig. 1.8 – NASA APOLLO Remote Site Data Processor – block diagram.

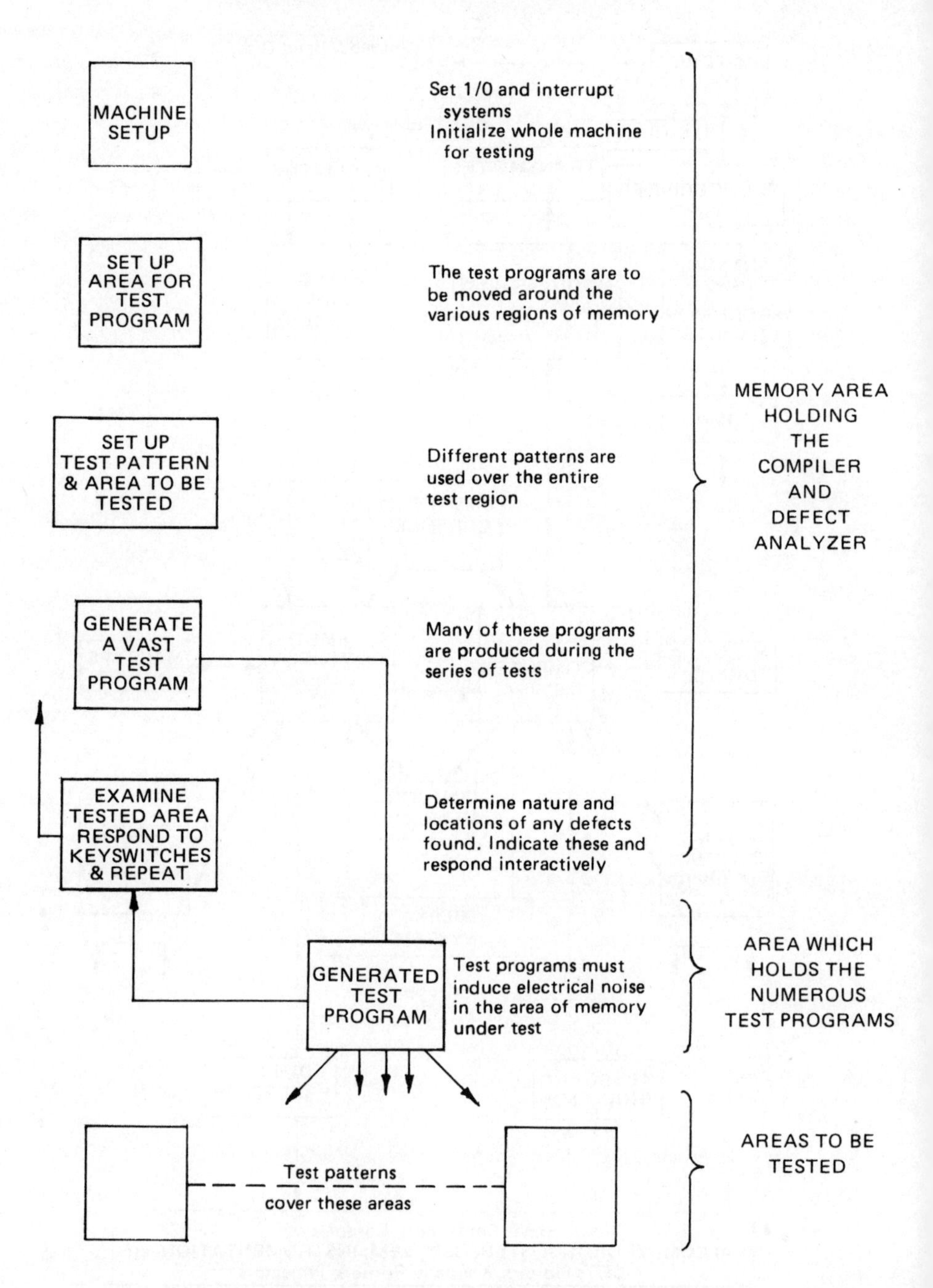

Fig. 1.9 – A generative memory-test program for a computer.

When the author took charge of the Carnarvon RSDP the MTBF of the system was of the order of a couple of hours. MTTR was some 15-30 minutes. A very strong team of professional engineers and technicians worked uncountable hours to achieve an eventual MTBF of two weeks for the whole system which represented more than a year for individual computers and other major items of equipment. The impending faults were deliberately induced to occur at regular servicing periods. New computer programming techniques were devised so that an equipment released from maintenance was known to be in a truly serviceable state. An early problem with computer test and 'diagnostic' programs was that they would run very satisfactorily in computers with certain classes of marginal defect. The new style of test program (Fig. 1.9) was designed to induce sensitivity to noise and marginality in the computers and hence trap impending fault conditions.

Some of the ideas from the previous pages can be summarized by relating complexity to usefulness and usefulness to availability via a series of approximations so as to deduce trends in system performance over the 50-year span.

The availability of a system for operational use is dependent on MTBF and MTTR. If MTBF is taken against 'real' dates, not related to downtime, then:

$$\text{availability} = 1 - \text{MTTR/MTBF}$$

However, MTTR is commonly very low, being simply a matter of minor system procedures to 'bypass' the defective part or to reload the affected computer.

Unfortunately, the situation is complicated by the fact that many systems must perform against schedules of high activity. We commonly find that the published availability figures (often 95-98% is claimed) fall to around a real 50% during scheduled high-activity periods. The quiescent system often performs better than the fully-active system.

Any attempt to derive formulae for availability and usefulness is highly approximate and one searches for improvements measurable in order-of-magnitude. Accordingly, for the present review, the term 'utility' has been adopted and defined as:

$$\text{utility} = \text{component count} \times \text{MTBF (in days)}$$

1.2 IMPROVEMENTS IN PERFORMANCE DURING 1930 to 1980

Electronic systems during the early 1930s were often confined to the use of a small number of thermionic devices together with relatively simple capacitors, resistors, inductors, switches and interconnections. A typical 'small' system would contain some ten or so 'components' and require attention every three months. A figure of number-of-components times MTBF in days was of the order of 10^3. Larger systems tended to exhibit shorter MTBF – and often longer MTTR so that availability tended to fall. However, the increased complexity gave higher values for usefulness and there tended to be a comparable 'utility factor' between small and larger systems.

By the 1940s, equipments containing some hundreds of components could survive three months giving a utility measure of some 10^4. In fact there appears a trend to increased complexity, reliability and utility at around ten times per decade (Fig. 1.10) until by 1980 we saw systems of 10^5 components lasting some years as evidenced by certain deep-space probes. Present-day values of 10^8 are quite common for system utility. Some of the techniques for these changes will now be reviewed.

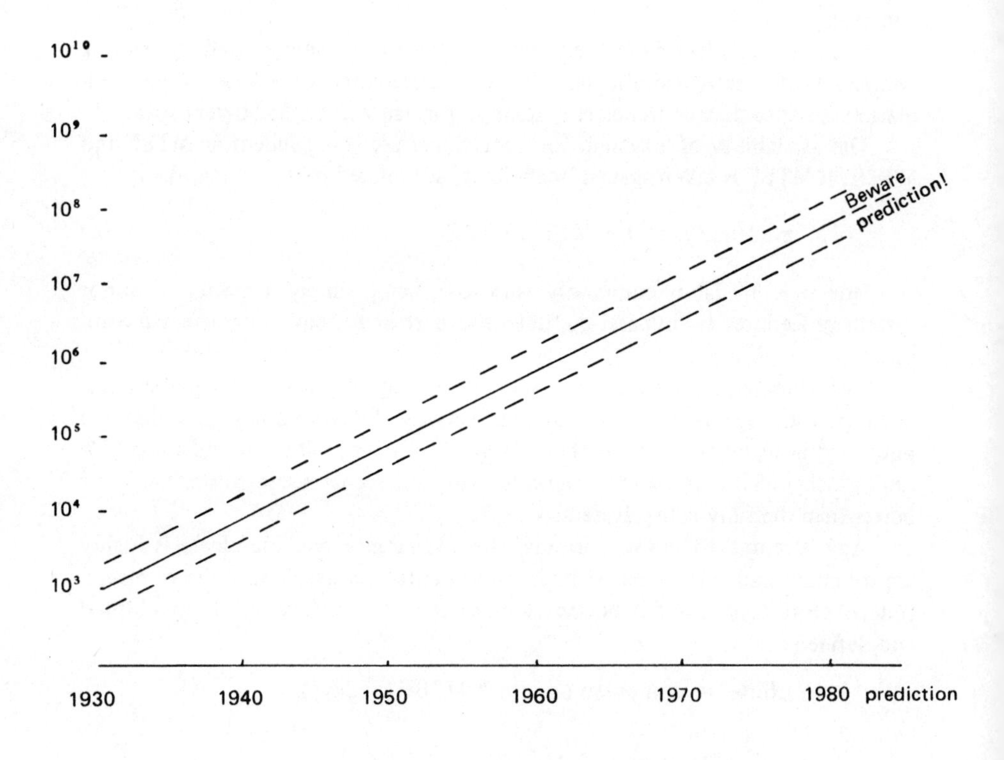

Fig. 1.10 – Components × MTBF 1930-to-1980 and beyond.

Figure 1.10 illustrates the trend in utility of systems over the 50-year period. However, beware extrapolation. Given that the data are 'noisy' (the normal case), no known mathematics exists by which the commonly used Linear Regression can be proved valid in extrapolation.

1.3 IMPROVEMENTS IN COMPONENTS – THE CONCEPT OF THE COMPONENT

References to 'components' in the foregoing paragraphs have been excessively simplistic. Consider the old-fashioned component shown in Fig. 1.11, the 'carbon rod resistor', which held sway for many years. A component? It had a number of constituent parts: the compound carbon rod of a mix and dimensions which would result in the required Ohmic value and power dissipation capability. To the rod were affixed two tinned copper wires – the mechanical and electrical contact being made using solder. This simple component was a complex structure with a number of potential failure modes.

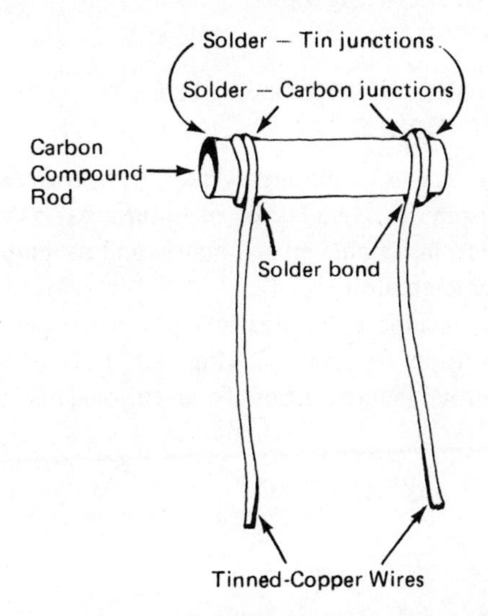

Fig. 1.11 – Carbon Resistor – a component?

Again a simple control – say a variable capacitor – has many mechanico-electrical junctions, any of which may fail. Some of these are extremely prone to failure in one of a variety of modes.

The very atmosphere in which a component operates becomes a functional part of the component, normally enhancing potential failure.

Atmospherically-sealed components have provided considerable improvements in reliability. Advances in the chemistry and physics of materials have played dramatic roles in improving system reliability and have in fact, made possible the construction of many complex systems.

1.4 IMPROVEMENTS IN INTERCONNECTIONS

The 'dry-soldered joint' plagued electronics for many years. There was a computer which, prior to powering-up every morning had to have up to a hundred joints resoldered. The soldered joint was seriously affected by chemical impurities on the soldered surfaces and atmospheric effects could accelerate deterioration. There was a rather famous PCM demodulator of the 'random logic' form of design in which the continual blast of cooling air caused a crystallization of many thousands of soldered joints.

Interconnections have been made using screwed terminals and plug/socket matings. The 'wire-wrap' technique solved many of the problems – only to introduce its own brand of defect and difficulty.

The use of non-corrosive surfaces such as gold has helped greatly in improving reliability of mating surfaces but even so, it is common to find connectors being remated regularly as part of a servicing routine.

1.5 REDUCTION IN INTERCONNECTIONS – THE LSI MICROELECTRONIC DEVICE

By fabricating a complete device (components and interconnections) in a single material unit, the atmospheric degradation of interfaces is minimized. All components are maintained in a stable crystal matrix and so temperature effects are constant across the complete unit.

The microfabrication technique ensures low power consumption and constant environmental conditions for the unit (Fig. 1.12). These factors remove many of the previously-normal failure centres of conventional discrete-component designs.

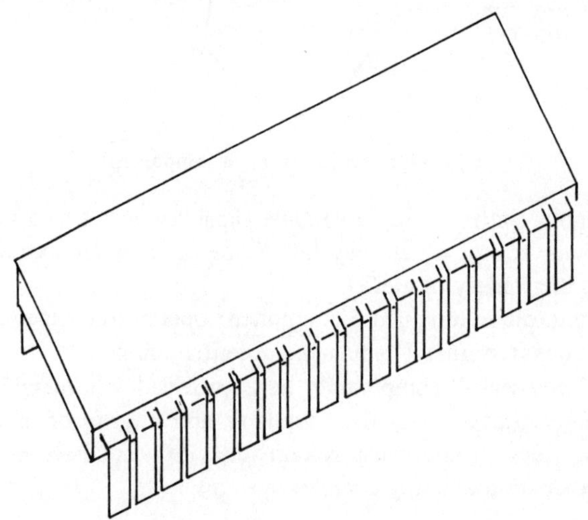

Fig. 1.12 – A microprocessor of 50,000 transistors – a component?

In a decade, we have moved from the computer having many thousands of components and vast interconnection fields to the computer which is a single compound component. The effect on reliability is little short of staggering. We can now confidently design equipments containing hundreds of computers, knowing that the system will exhibit hitherto unheard-of reliability.

1.6 FAILURE MODES IN COMPONENTS

1.6.1 Degradation

Our old-fashioned thermionic tubes suffered gradual reduction in electron emission from the active surface(s) and occasionally reduction in vacuum level by imperfect sealing of metal-to-glass. Resistors would alter in Ohmic value due to changes in the conductivity of the metal:carbon bonds. Such effects are commonly dependent upon the temperature of operation. In many cases the degradation was self-perpetuating. The famous 'thermistor' would exhibit an accelerating degradation when used in a current-carrying capacity. A lowering of conductivity either of the base material or of the the metallic connections could result in increased heat generation within the thermistor. Because the degradation was enhanced by thermal effects, once conductivity fell below the design rating for the application, the degradation would accelerate. However, the onset of degraded performance could frequently give rise to very puzzling system defects.

The principal electrical evidences of degraded performance of a component would be either reduced conductivity in devices which should conduct or the onset of conduction in devices which should not conduct electric current. When a device is to handle alternating currents, the degradation effects become more complex leading to considerable difficulties in testing. Capacitors are connected into circuits by inductive and resistive wires. Inductors have internal capacitance and virtually any of the parameters may be subject to deterioration.

1.6.2 Solid Failure

Certain total failures of components can result in failure of an equipment to perform its required function. Other component failures can result in degraded performance which is obvious and readily-traceable. Yet other component failures may go unnoticed until a full calibration check of the equipment is made.

In general, the solid failure is the most straightforward type of component defect to trace.

1.6.3 Intermittent Failure

The most troublesome of all forms of component failure is the 'intermittent' defect. Intermittency is commonly due to a mechanical imperfection which causes a variation in electrical conductivity. The defect may not evidence itself until some mechanical disturbance occurs. Vibration due to a loudspeaker can disclose defects in components of the electronic devices and circuits driving the speaker. The servicing toolkit has been known to include 'soft' hammers for tapping components. Even a two-faced hammer with one hard-rubber and one soft face has been found as part of the trouble-shooting equipment. If the tap with the soft face discloses a defect, we have a problem. If a tap with the hard face causes no effect, its 'all systems go'.

If we treat all potential failures in components as falling into the category of intermittent defects, we can develop theoretical approaches to system testing which result in the development of highly effective fault-analysis procedures.

1.7 EFFECTS OF COMPONENT FAILURES

1.7.1 Effects of Failures in Analog Equipment

Component failures of the three principal modes can cause:

(a) excessive 'noise' with all that that can lead to
(b) degraded performance or excessive signal distortion
(c) servo hunting and other forms of resonance effects
(d) total or intermittent failure.

In many cases, minor perturbations pass unnoticed or create but passing interest – until major servicing is due. But when a defect causes a disconnection, such as the 'dirty contact' which unlatches a relay for example, it is time to take action.

A lack of calibration in a navigation system could have very serious consequences and hence such equipment must be designed and built with great care – and subjected to regular servicing checks.

1.7.2 Effects of Failures in Digital Equipment

Component failures here provide quite different effects than are those related to analog equipment. There are very few strictly digital electronic components – almost all are analog components designed to behave as though they were actual digital devices. The simple 'logic gate' is constructed from analog amplifiers with 'clamping' circuits such that an effective high-rate transition between two preset voltage or current levels occurs. The effect of gain degradation of these amplifiers is to reduce the slew-rate and hence increase transition time between the preset levels. This can result in delaying the onset of a transition and may cause malfunction of a subsequent gate in the logic chain.

Probably the most difficult of all defects to trace is the intermittent failure in certain parts of a digital computer. Many (often most) logic elements of a computer are rarely used during operation. It can then happen that a defect strikes a circuit at a time when it has no part to play in the current operation – only when the defect strikes and that actual circuit is in use could the effect be noticed: even then, the defect could cause the defective element to be in the 'wanted' state so that no defect is observable.

An intermittent defect in a memory cell for instance, may cause incorrect information to be written to that cell and this may go unnoticed for a considerable time. One such failure almost resulted in the loss of a manned spacecraft. Only precise navigation saved the APOLLO 10 spacecraft when a computer failed during transmission of the 're-entry commands'. The defect had been under investigation by a vast support team for more than a week. When the defect 'went solid' it took just a couple of minutes to trace the offending PCB. However, another three months was to pass before 'the lab' determined the cause – the nightmare of a 'hairline crack' in a copper/gold trace.

How, in a complex computer system, do we set about the first stage in fault-finding – tag it as a hardware or as a software defect? Certainly the hardware is Analog in nature. Only the software can be truly two-state. In system design we must insist that 'software bugs are not permitted'. Impossible? Essential! And achievable in practice. Once we have systems in which this can be accepted as fact, we know that any defect is hardware-induced. So our hardware must be verifiable in every mode of use. Designs containing large numbers of computers can now be contemplated with confidence – using single-chip computers and other LSI components. But we must include system validation tests in the design.

1.8 TECHNIQUES OF SYSTEM OPERATION

How should you treat an electronic system? All may be well until a component fails – but do you switch off to effect a repair? If you normally leave the equipment powered-up, when the repair is made and you restore full power the next couple of weeks could be disastrous. The reason for this is an effect known as **thermal cycling**. Some of those interconnections and even some of those 'active elements' react poorly to the cooling and reheating cycle – so another failure. It is quite common to find that two weeks elapse before the system finally settles down to reliable service following a repair.

Should you then deliberately thermally-cycle the equipment, say daily, or weekly. This will induce more defects at the onset – but can result in a system which is operationally more stable. Potential failures are almost deliberately induced at a scheduled point in time so that repair can be effected before operational use is disrupted.

Oh yes – and how do you test a computer? There are just too many combinations of logic states to be tested within a reasonable life expectancy of the machine. Further it is well known that computer test programs run rather well in defective computers – because they use the computers in modes never tried by normal operational programs. Theoretical studies have disclosed that we 'cannot write test programs'. Valid test programs can however be generated within a computer. The concept of the 'validation test' has taken a long time to filter into the world of the computer. Test programs must treat the computer in a manner even more devastating than can 'operational' programs.

1.9 USE OF REDUNDANCY IN SYSTEM DESIGN

The processes which have led to the dramatic improvements in component and system reliability can be pushed just so far. We may predict that by 1990 we shall see systems with 100 million 'components' surviving the magic three months in the form of vast arrays of extremely complex 'chips'. But we can expect the designers of the '90s to be searching for still higher values of utility and reliability.

However, when safety is paramount, few of us would care to rely on a system not failing through simple high-reliability. If a defect should occur just as the autolanding system is carrying ME back to ground .... The prospect is just too much.

So, we install two or three copies of each critical module and provide a switching mechanism such as shown in Fig. 1.13. This results in a considerable number of alternate signal paths through the system. Some paths may be selected automatically whilst others require manual intervention. Harking back to the NASA APOLLO system, there were numerous around-the-world communications pathways and it was normal that the 'birdie' was tracked by two or even three ground stations – with others available to take over in an emergency. Within a tracking station, some idea of the degree of redundancy available can be gained from Fig. 1.8, noting the symmetry of the computer installation. A vast bank of multipole, multiposition switches was used to determine the actual routing of each major assembly (and each computer channel).

1.10 USE OF AUTO-TEST CIRCUITRY

Certain types of system can – and should – be provided with automatic testing facilities so that, even during operational use, indications of unsatisfactory performance can be provided. In some cases, it is possible for automatic corrective action to be initiated; in other cases, manual intervention is required. A number of the sections of a computer can be designed with autotest and even auto-corrective facilities incorporated. Internal memory systems can include additional 'parity' bits. By careful use of coding theory, sets of such bits can be used not only to indicate a corrupted bit in memory, but also to correct the defective bit so that no loss of information occurs. However, such auto-corrective action can be reported by the computer so that the maintenance team has diagnostic records.

An autotest circuit failed during an APOLLO manned spaceflight: the computer was instantly disabled by the protection circuitry. The 'second copy' computer was unable to take over due to the nature of the defect. Amidst loud protestation from the Flight Control team, the 'Battle Short' switch (they used modified military computers for APOLLO) was thrown to disable the autotest circuits – and mission support continued.

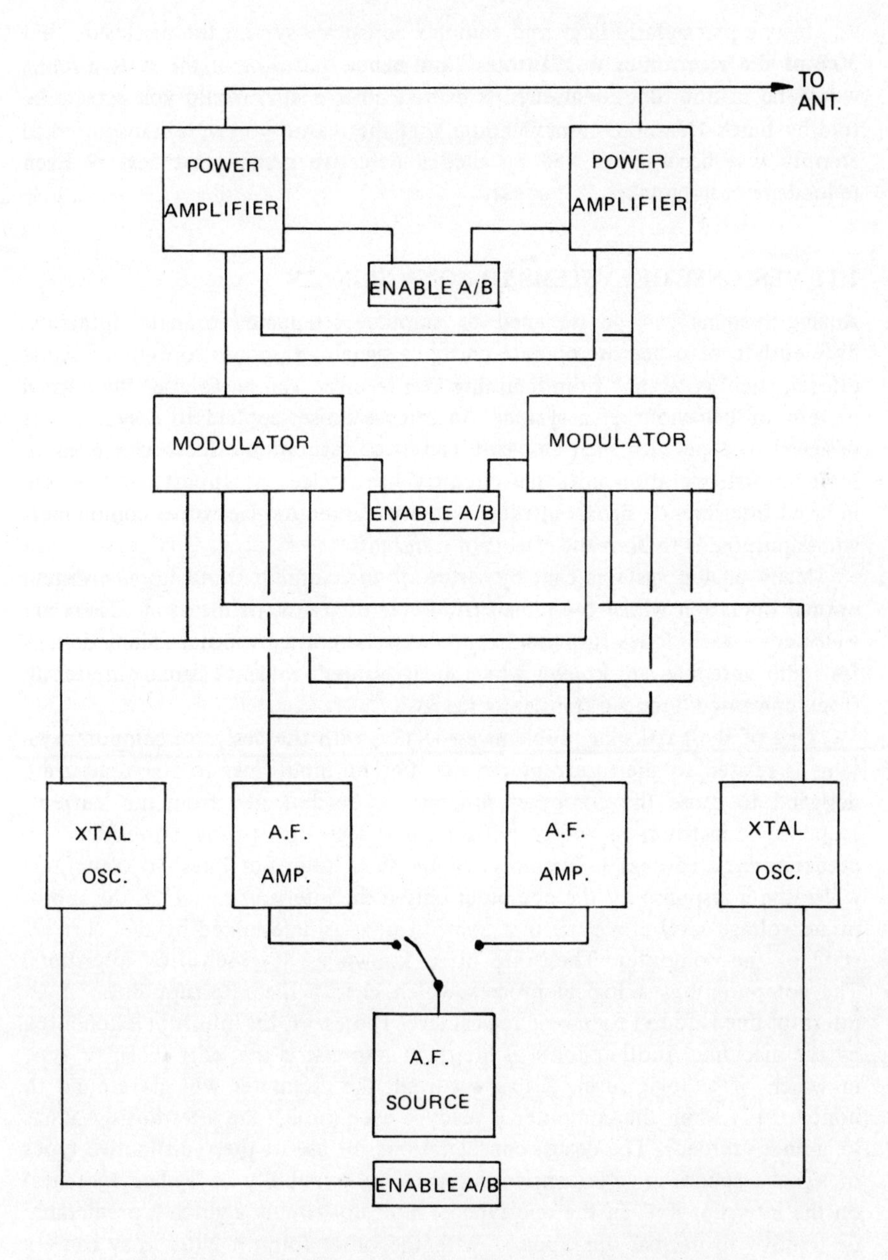

Fig. 1.13 – Redundancy in a radio transmitter – twinned modules.

In one particularly large and complex aerospace system the disclosure that 30% of the electronics was 'autotest' and hence that 30% of the system faults would be in non-functional circuits caused quite a stir. Would you care to be told by Earth–Heaven communications that the reason you were transported to eternity was because we had installed a defective system test device? Even redundancy can be taken just so far.

1.11 RESPONSE OF SYSTEMS TO NOISY SIGNALS

Analog systems can be designed to amplify, attenuate, resonate, integrate, differentiate or otherwise operate on input signals. Response to extreme signal effects, such as 'static' from lightning in a receiver, can be to alter the normal pattern of behaviour of a system. An intense pulse, applied to a system not designed to cope with such an event, can cause saturation effects which can inhibit normal operation until the circuitry has settled. At circuit points where induced interference signals intrude, we have learned to place other components whose purpose is to limit the effects of transients.

Many analog systems can, by virtue of their inbuilt integrators, maintain normal operation whilst the remanent effects of transients die away. There are well known techniques for minimization of noise borne by signals. Many designs for radio antennae are known which treat wanted 'radiant' signals differently from 'unwanted' induced transient streams.

One of the particular problems associated with the design of computer systems is related to the interrupt circuits. Certain input lines to a computer are designed to cause the computer program to be deflected from the 'current' sequence of instructions and to initiate a new segment of program residing at a predetermined address in memory. Some such 'interrupt lines' to computers will cause a response by the computer only if the interrupt signal has the appropriate voltage level at a particular point in time as determined by the 'current' state of the computer. These are often known as 'level-sensitive' interrupts. The computer has a logical process which checks the interrupt lines: if an interrupt line is found to have a 'logical level 1' present, the interrupt is honoured by the machine. Another form of interrupt response is the 'edge-sensitive' type in which, if a 'logic level 1' has occurred, the computer will get around to honouring it when the computer is ready – even though the interrupt signal has long-since 'zeroed'. The design considerations for use of these distinctive types of signal response include considerations of the possibility of 'spikes' (glitches) on the interrupt line. In the one type of interrupt circuit, a glitch is predictably destructive to normal operation – with the other form, a glitch may cause a false interrupt. With the complex interrupt circuitry of modern multiple computer systems, great attention must be paid to the purity of signal lines which carry control signals.

Most digital devices are relatively immune to noise on the signal and control lines – but one all too frequently meets devices capable of inducing very large impulses. Such a problem in the NASA APOLLO computer system resulted in glitches of over 100 volts amplitude on signal lines with a normal 0–5 volt response.

1.12 PROCESSES AND PROCESSORS

In modern electronic design, there are many classes of 'signal' to deal with and many types of process to which signals may be subjected. We have a choice between analog and digital forms of processor and often mix the two in hybrid designs. Knowing the principal characteristics of the two processor systems, we are able to decide which processes are best dealt with by analog and which by digital processors.

Decisions regarding the types of processors to use at different stages in the processing task have a considerable bearing on the ultimate reliability of the completed system. The decision on use of analog or digital process has varied with the years. At one time, only elementary control systems and telephone exchanges operated on digital switching. With the development of computers and of the microelectronic versions of computers, we now conciously decide to move many processes from the continuous world of analog processing to the discrete domain of the digital system.

We frequently select the digital processor for its greater long-term stability of systems parameters and accept the possibility that the added complexity may create penalties in reliability. The general trend however, is towards the use of the LSI microelectronic device which offers both increased utility and reliability in systems.

1.13 DISTRIBUTED-FUNCTION PROCESSORS

Now that we are able to construct multiple-computer systems with the ease provided by microelectronic systems, we can break a complex computer program into its 'processor' parts and allocate different processes of the program to different computers. This enables us to adapt the high reliability of relatively simple computer programs to the fabrication of complex systems.

The conventional 'time-shared' computer is commonly used in specific ways to perform a variety of tasks. By identifying these tasks, we can identify new approaches to computer system design based on the concept of 'task processors'. A computer user needs access to a set of common services which are best provided not by a single processor but by many processors coupled to common services such as libraries of specialized computer programs.

The move then is toward distributed processing and to vast task-processor computers. The very large machines should be used for what they are best suited – not to service the keyboards of many potential users.

Vast quantities of information are conveyed via parallel, serial, discrete, and carrier systems using varieties of modulation and coding systems. Defects in any part of such systems can cause corruption of the transmitted information. Error-detecting codes can be used in certain circumstances – but often, the reduction in quantity of information transmittable dictates that only superficial checks be made. Great reliance is placed on high-quality systems and reliable communications paths.

1.14 LIMITS TO COMPLEXITY – AVAILABILITY

Present-day trends increase the utility of systems by great increase in complexity and by the linking of many systems. The problems of the common carrier have increased since the introduction of stored-message communication. If a sender's message can be accepted and stored until the local station can transmit the message, there is a considerable improvement in utility for the user. Store-and-forward has been in use for many years. Because of its success, we now require the transmission of million-character messages. Store-and-forward enables the communications system to handle corrupted transmissions by message validation (error-correcting coding and retransmission). Now, our communication channels can introduce errors but correct them even without our knowledge. What if somewhere in our multistation communication network, corruption is causing excessive retransmission – can we even know that this is happening? Could our communication system grind to a halt with vast stores of in-transmission messages all being continually retransmitted? There are many cases of this situation arising during the development of such systems.

We speak of the 'logical length' (or depth) of a signal path and now find systems with almost uncountable quantities of logic elements and interconnections. Are we nearing the limit of controllable complexity in communication and processing systems?

Although utility increases with complexity, availability falls due to increased MTBF. We introduce redundancy, star- and ring-networks adding to complexity and to the very important MTTR.

1.15 RELIABILITY VARIATION DURING OPERATIONAL LIFE

Whilst there are very considerable improvements being made in the reliability and performance of equipment, it is generally found that a new installation suffers a number of defects in the initial stages of system validation and operation. As time goes by, the situation tends to improve, both as marginally defective components disclose their whereabouts and as servicing routines become firmly established.

There can be periods of unsatisfactory operation during the main operational lifetime of the installation due to thermal cycling effects as discussed earlier and to the tendency for maintenance to become a mundane matter – there being far more important things for people to do.

However, time proves not to be the great healer and, after some years of operational life, the MTBF tends to decrease. There comes a point when economics, operational requirements, and the appearance of new equipment of higher operational utility dictate that the 'end of life' point has been reached. This is usually expressed in the now-famous 'bathtub curve' of Fig. 1.14. No scale is provided on the illustration of the bathtub – so much depends on the nature, purpose and modularity of construction used in the equipment. Some systems may expect an operational life of a few years, others may soldier on towards the 20-year mark.

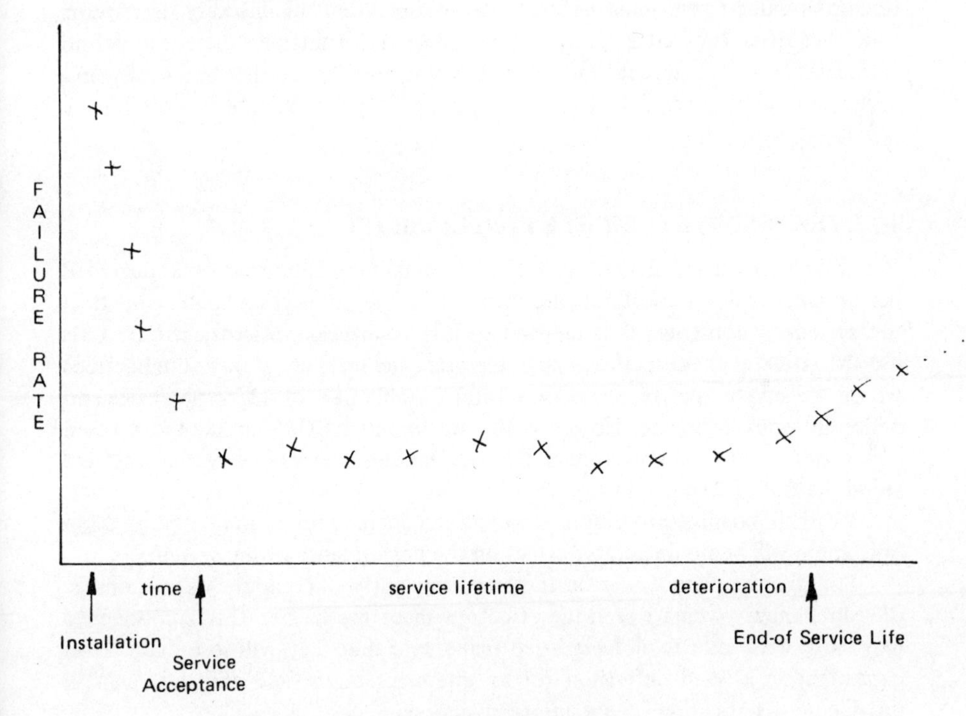

Fig. 1.14 – The Bathtub curve.

1.16 NUMBERS INFORMATION, MEASURE AND MEASUREMENT

Information is represented in many forms in electronic processor systems. There are however some very perplexing problems in the use of numbers for information representation. We have applied the theory of mathematical induction – counting – and derived theories of countability. We are able to classify countable and noncountable sets. We can define many forms of arithmetical and mathematical operation upon numbers and sets of numbers. We call it computing.

However, computing carries many curious effects. Not only do we need to avoid the possibility of dividing by zero – we also have to be aware at all times that changes of radix pose problems. Most measurable values cannot be represented in any known radix. Because of this, all measured values are stored and processed with limited precision. Even the sequence in which relatively simple arithmetical operations are conducted can affect the result significantly. A number of computer programming languages have failed due to yet another characteristic of the digital computer: not only must we know how to handle numbers in differing radices, we must also know how to handle the representations of numbers. We must define a set of operations on the very descriptors which are often implicit in computer programs. unfortunately, there is as yet no mathematics for the processing of such descriptors, hence the failure of many programming languages. Perhaps 'Computing Without Numbers' would be a safer thing to aim for.

1.17 THE POSITIVE PRINCIPLE IN RELIABILITY

If a problem is worth solving – avoid it. If a component failure can cause degraded performance – don't use that component. If an interconnection failure can affect the system – don't use that method of interconnection. Microelectronic LSIs use the latter of these methods by integrating the inter-component connections within the crystal matrix; there they form a stable part of the single device and hence are unlikely to fail. However, the old fuse-link ROMs could regrow fused connections and result in system failure. Has the present-day technology yet stood the test of time?

We shall examine processors in which, if a component or an interconnection fails, there will be no detectable effect on the performance of the system.

For this, we cannot rely on increased reliability of components, nor on use of redundancy – we have seen the effects of these approaches. Our 'components' may 'lose data' due to defects – so make *sure* that they will lose data. If we depend upon a fixed definition for an interconnection field and this can fail, provide *no* specification for the interconnections.

Use unreliable components – do not use known coding systems – have no 'wiring list'. If we can make the thing work at all, we can allow all manner of damage to occur – and it cannot fail to continue in operation. Impossible? Perfectly possible – as we shall see.

2

Geometrical pulsator matrices – fundamental theory

2.1 THE SEARCH COMMENCES

In this chapter we shall commence the search for systems which can offer improved reliability. The chosen start-point is a remarkably simple electronic circuit which is rather rich in feedback and can exhibit somewhat unusual properties. The problems of analysis open the way for attempted applications of similar circuits and to different and more satisfactory analytic techniques.

2.2 THE AND-END ELEMENT

The building-block for the initial circuit is the rather outdated AND-END device. In a system we define time-slots of equal duration and a 'standard pulse' which occupies one time-slot. The function of the AND-END device is that, if coincident pulses appear at the inputs the device will produce a standard pulse output filling the subsequent time-slot. Figure 2.1 illustrates the relationships between inputs and output in time-amplitude form. From Fig. 2.1, the pulse at the output during time-slot 8 resulted from the termination of the co-incidence of pulses at the A & B input during time-slot 7. Other inout states cause no output.

The effect of the AND-END element can be summarized as:

$$\text{Output} = d(\text{A} \,\&\, \text{B}) = \text{in 'delay' terminology}$$

or as: $$\text{Output}_{tn+1} = \text{A}_{tn} \,\&\, \text{B}_{tn} \text{ in time sequence.}$$

2.3 PULSORCUBES

The arrangement of AND-END elements which we shall examine is based on a cubic form. We shall consider that one element occupies each apex of a cube

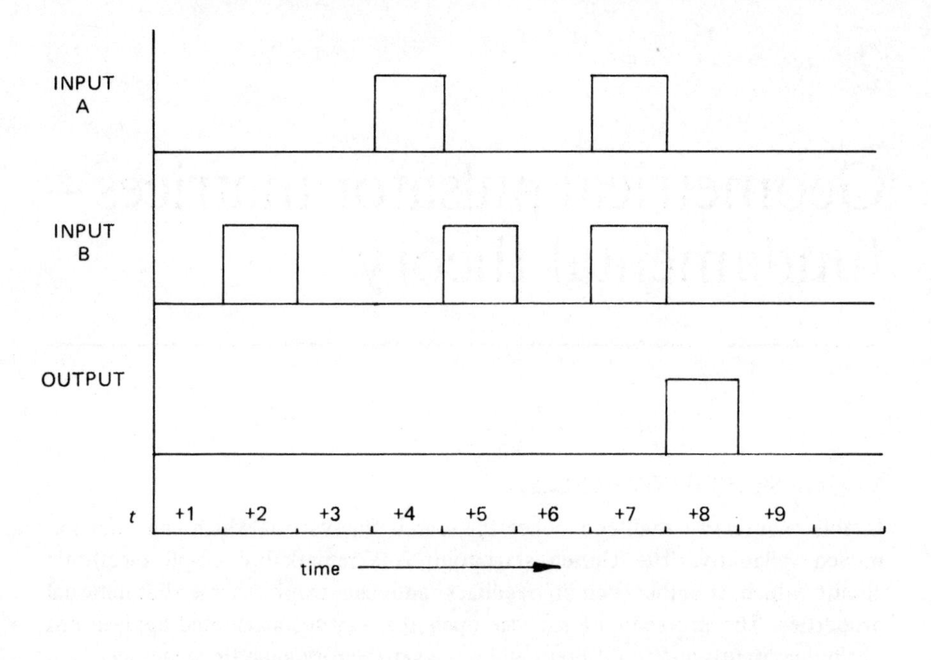

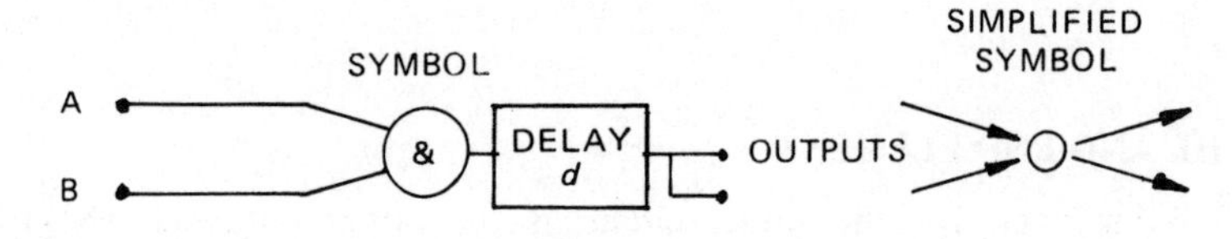

Fig. 2.1 – The AND-END element.

making a total of eight active elements, as outlined in Fig. 2.2. We shall not consider the details of the manner by which system synchronization would be achieved were we to construct the device; we are concerned with the possible functioning and behaviour of such a device. It is a stepping-stone to more realistic models.

2.4 PULSORCUBE INTERCONNECTIONS

The functional interconnections between the various elements of our device commence as illustrated in Fig. 2.3(a) where the upper plane of the cube is treated. Connections are established in a progressive input-to-output manner along two parallel paths. One input is made to the 'left-hand' element and one output is to be taken from the 'right-hand' element.

The lower plane of the cube is used in a similar manner to establish a second parallel pathway with its own input points as indicated in Fig. 2.3(b).

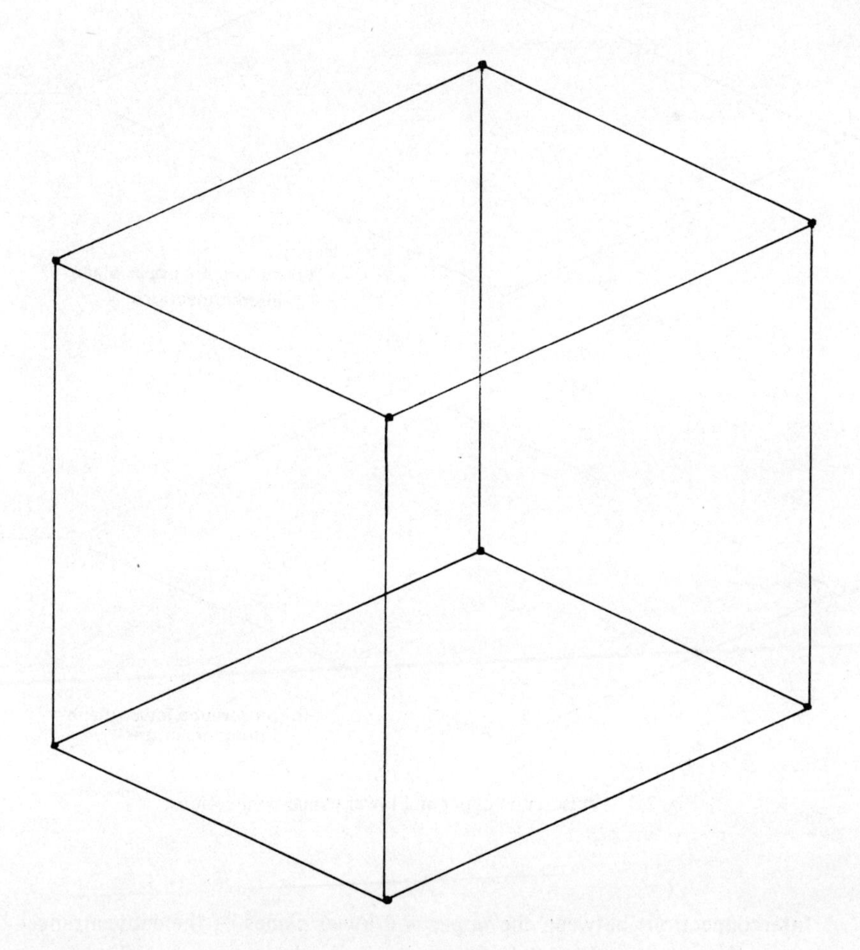

Fig. 2.2 – The basis of the pulsorcube.

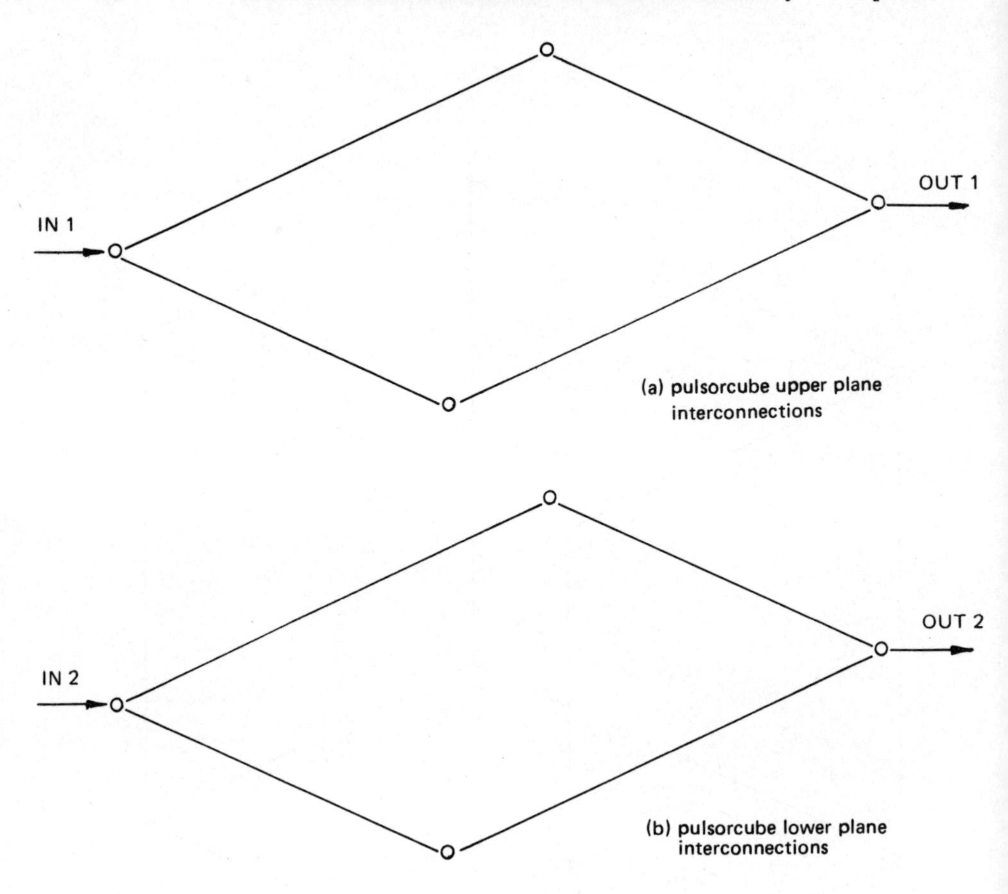

Fig. 2.3 – Pulsorcube upper and lower planar connections.

Interconnections between the upper and lower planes of the cubic arrangement are illustrated in Fig. 2.4. Once a start has been made on this circuit design, it is found to be quite systematic. Noting that as a result of the upper and lower plane connections, and that each element has two input and two output connection points, it may be seen that the output points of the left-hand (input) elements are fully utilized and that each of these elements requires just one further input line.

Cross-connection on the diagonals of the 'left-vertical' plane interconnect the upper and lower planes in a reflex manner. These interconnections satisfy the terminal requirements of both the input elements. We make similar interconnections at the opposite face of the cube thus satisfying the terminal requirements of the two output elements.

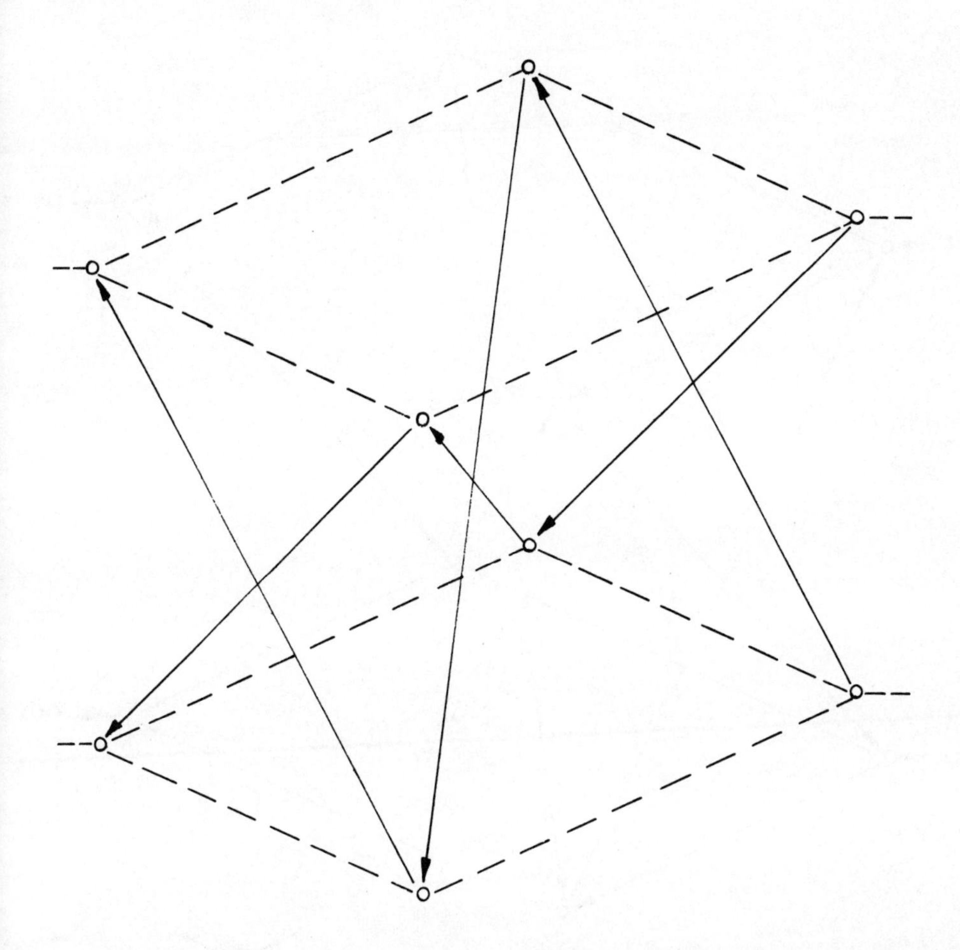

Fig. 2.4 – Pulsorcube – interplane connections.

There remain certain unsatisfied terminals at the intermediately placed elements. An examination of these terminals suggests a number of ways in which the remaining interconnections could be made. If the resulting device is to be truly rich in feedback connections, the use of cubic diagonals would be preferred to planar diagonals, hence the central pair of lines in Fig. 2.4.

The completed set of elemental interconnections for the 'pulsorcube' as the arrangement will be known is given in Fig. 2.5. Here the apices are annotated with letter references to the elements so that a logical analysis of the possible performance of the circuit can be undertaken.

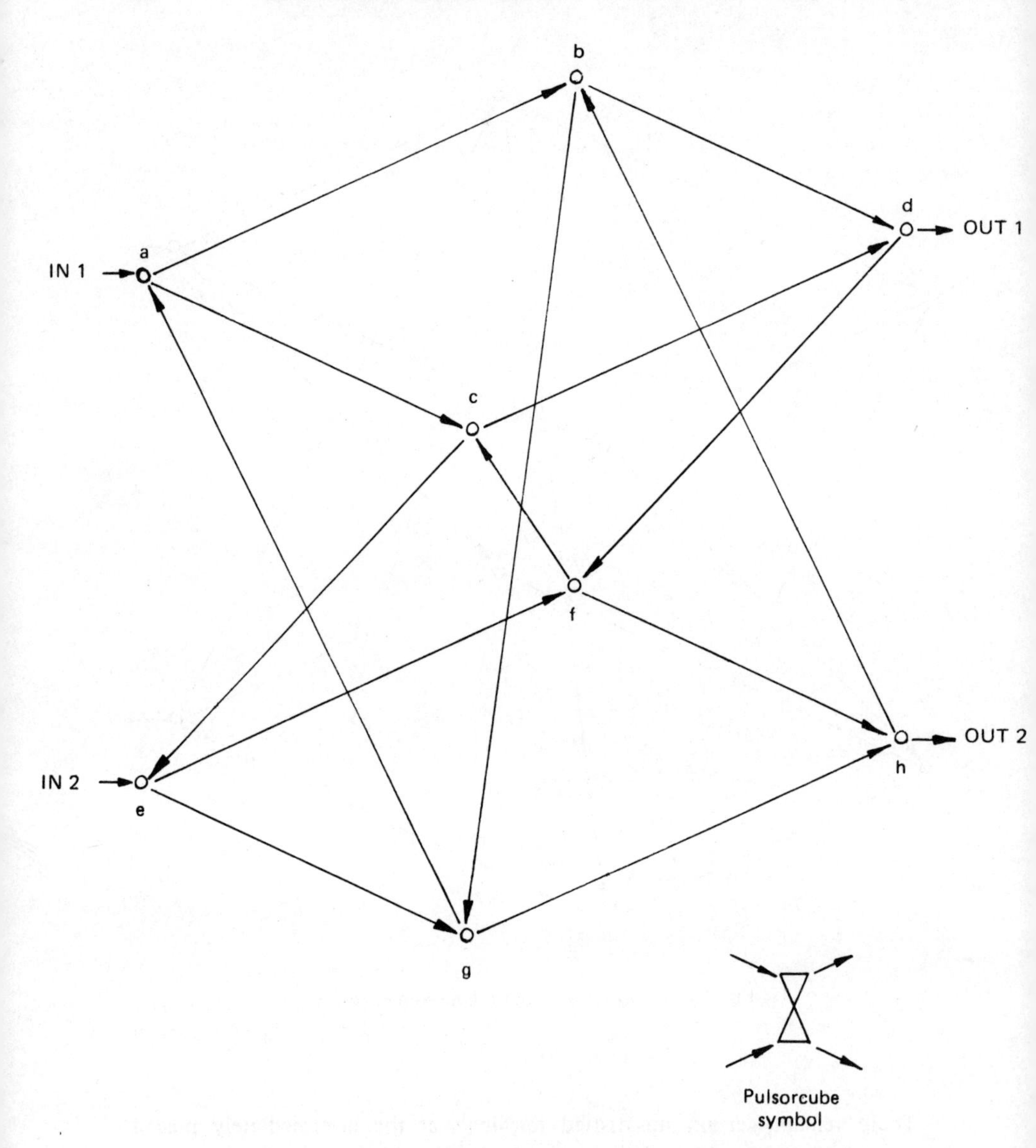

Fig. 2.5 – Pulsorcube – full interconnections.

2.5 ANALYSIS OF THE SYNCHRONOUS PUSLORCUBE

The output conditions for each element of Fig. 2.5. are given in Table 1.
Reference to Table 2.1 shows that, to produce an output at terminal 'd' (Fig. 2.5), inputs are needed from elements 'b' and 'c'. Element 'b' requires inputs from 'h' and 'a'. In turn, element 'h' requires inputs from 'f' and 'g'. These

inputs must occur at specific time-slots. Particularly, the input from 'f' must be at the time-slot 3 prior to the 'd' pulse. In the other path to 'd', element 'b' must be driven by elements 'a' and 'f'. However, this input from 'f' must occur two time-slots prior to that of 'd' pulse. This means that outputs are required from 'f' during two adjacent time-slots – clearly impossible.

Table 2.1
Conditions for pulsorcube output.

Element	will 'fire' after
a	IN1 & g
b	a & h
c	a & f
d	b & c
e	IN2 & c
f	e & d
g	e & b
h	f & g

By such reasoning then, we can see that the synchronous pulsorcube cannot deliver outputs.

2.6 ANALYSIS OF THE ASYNCHRONOUS PULSORCUBE

However, if we relax the requirements for synchronism and alter the ratio of input drive time to output pulse duration, quite new conditions arise. Let the requirements for driving a pulsator element be that the input '&' condition need occur for only one-tenth of the duration of a standard pulse – that the output pulse commences at the termination of the AND condition (non-synchronous AND-END). Attempts to analyse the operation of the pulsorcube will now fail for many reasons. One is that the sets of logical conditions leading to the initiation of a pulse are now of exceptional length; another is that, because we are using discrete circuit components and elements, we have no guarantee that precise timing requirements for analysis will be met. A practical asynchronous pulsorcube is non-analytic. Vast numbers of synchronous patterns can be supported (and 'memorized') and on input changes, the pulsorcube will transit from one stable state to another.

The circuit for an asynchronous AND-END element used in the fabrication of an asynchronous pulsorcube is given in Fig. 2.6. Such circuits are valid for both NPN and PNP transistors given suitable diode orientation. In the static condition, the high-valued resistor holds the transistor in a conductive state and

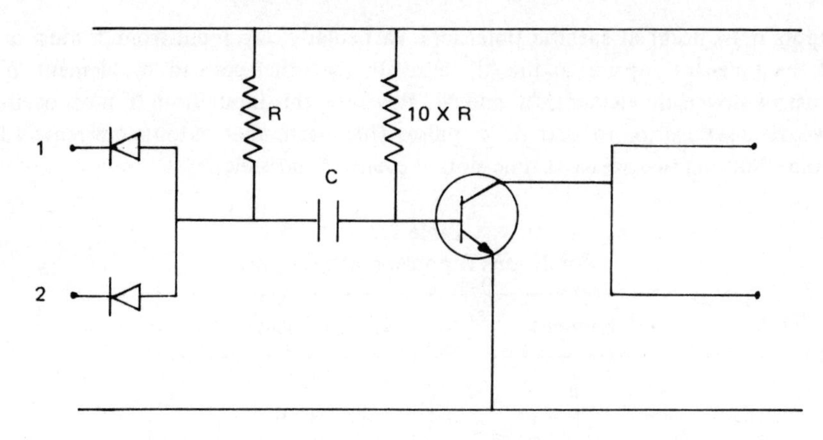

Fig. 2.6 – Basic circuit of an asynchronous AND-END element.

the output terminals are at near-ground potential. When the preceding circuits deliver 'high' inputs (+ve for NPN or –ve for PNP devices), the capacitor charges vir 'R'. Termination of the AND input condition, that is, either input returning to 'ground', causes the capacitor to cut off the input current to the transistor which then ceases to conduct. As the capacitor discharges, the transistor becomes conductive and the output is clamped at 'ground' once more. Figure 2.7 illustrates this point.

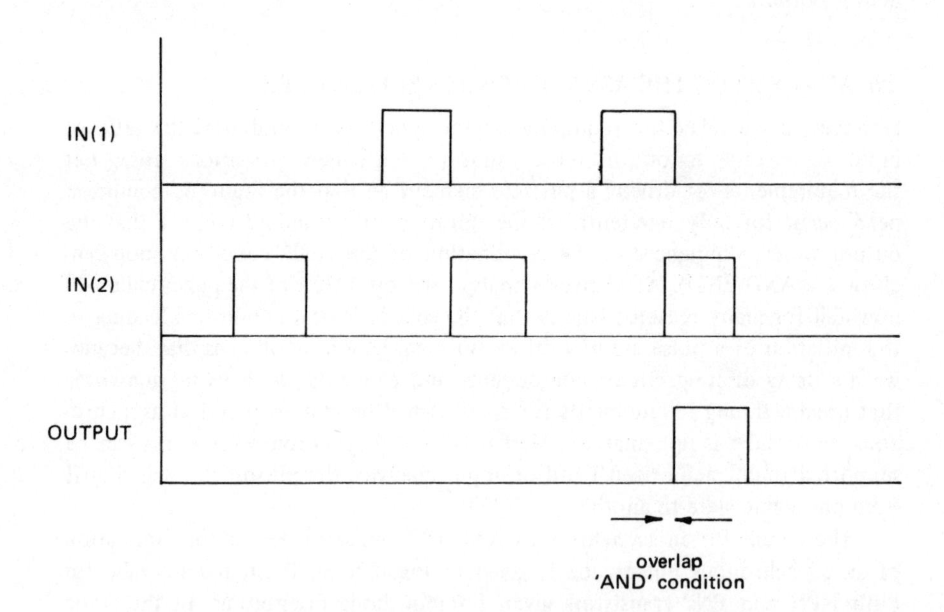

Fig. 2.7 – Asynchronous AND-END element – timing.

A small number of asynchronous pulsorcubes was constructed and their behaviour investigated. Results were a trifle unpredictable, certainly non-analytic by the mathematics at our disposal in the early 1960s – but they were such fascinating devices to toy with that, even though there seemed no practical place in the circuit books for these devices they received considerable attention in the limited circle of researchers involved.

One characteristic feature was a considerable number of stable states – dynamic states that is. At each node there would appear stable patterns of activity – waveforms. The number of such patterns was large – very large – unexpectedly large for so small a number of active elements. Here came an early clue which guided thinking in later work. With such complex feedback came 'memory'. Could it be that the number of elements does not determine the number of memory patterns?

2.7 THE SUPERPULSORCUBE

The pulsator element has two inputs and, in the example first considered, has an output fan-out of two, that is, two outputs. The pulsorcube is composed of pulsor elements and also has two inputs and two outputs. The superpulsorcube is simply a set of pulsorcubes considered situated at the corners of a cube and connected in the same manner as are the elements of a pulsorcube. The resulting circuit (Fig. 2.8) will have two inputs and two outputs. Only 64 elemental circuits are used in the Superpulsorcube yet the degree of complexity is, to say the least, rather interesting. Whilst the number of stable modes of an individual pulsorcube is considerable, for the superpulsorcube computation is not feasible.

The circuit is asynchronous so there are no time-slots into which special patterns can be mapped. Although the individual elements are 'two-state' devices, and each delivers a 'standard pulse', we have no means to calculate the instant in time at which any individual element will be triggered. The pulse streams within the device have an analog property in time. We cannot rely on using the 10% pulse overlap ratio as fundamental, slicing time into one-tenth pulse time intervals, because each element has its individual overlap sensitivity which will desynchronize the system.

The superpulsorcube remains an intriguing device which could be used to switch patterns of lamps in an interesting display – but it is a device which appears to offer no practical application.

2.8 THE PULSORHYPERCUBE

This device is another simple development from the concept of the pulsorcube. However, for the new circuit, our elemental devices have three inputs and three

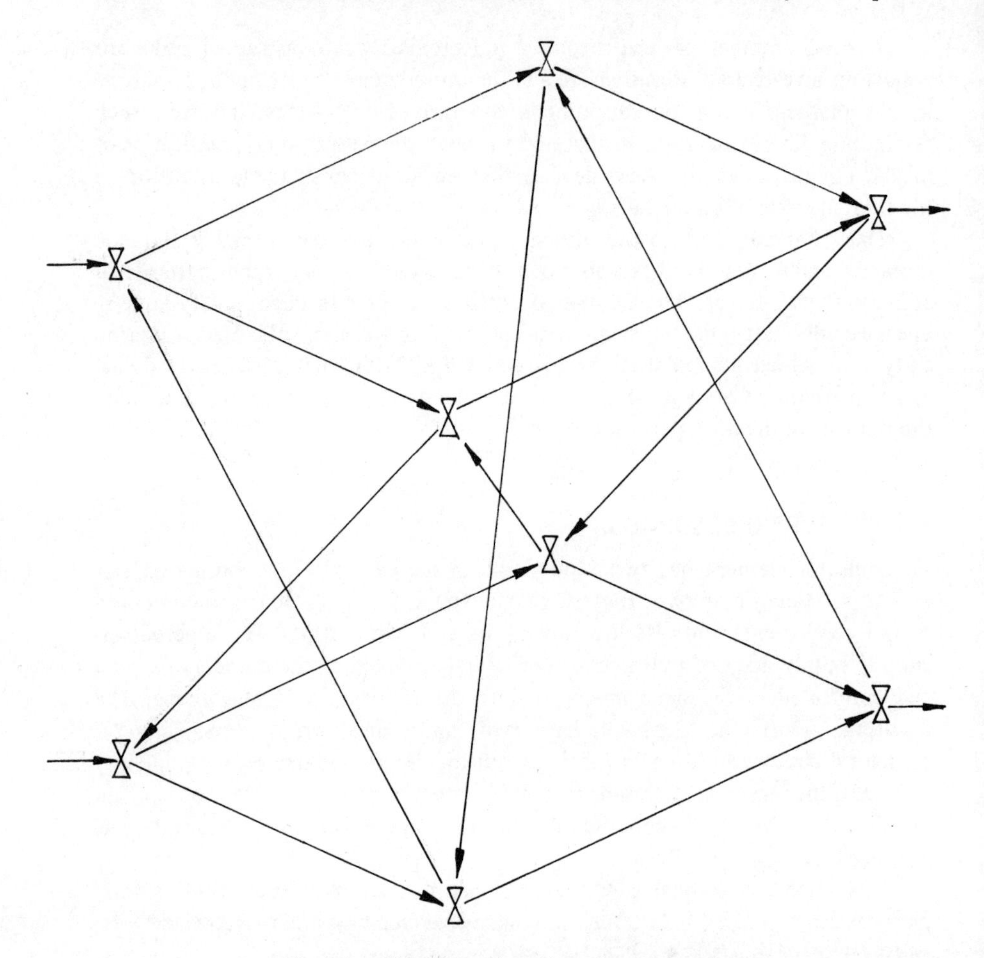

Fig. 2.8 – The superpulsorcube – interconnections.

outputs. Rather than position the elements at the apices of a cube, we use a tesseract or hypercube. One view of a hypercube frame is illustrated in Fig. 2.9.

The interconnections of the pulsorcube (Fig. 2.5) result in four connections at each apex of the cube – two inputs and two outputs. By providing a third input and a third output to each element, as shown in Fig. 2.10, we can satisfy an additional spacial direction – an additional orthogonal direction at each apex.

We are accustomed to the representation of three-dimensional objects on plane paper and have seen a number of views of solid objects. In three-space we can see only one solid view of a tesseract. A number of decorational objects have been based on views of tesseracts.

The principal claim to noteworthiness of the pulsorhypercube is that it is a four-dimensional device which can be fully constructed electrically and any

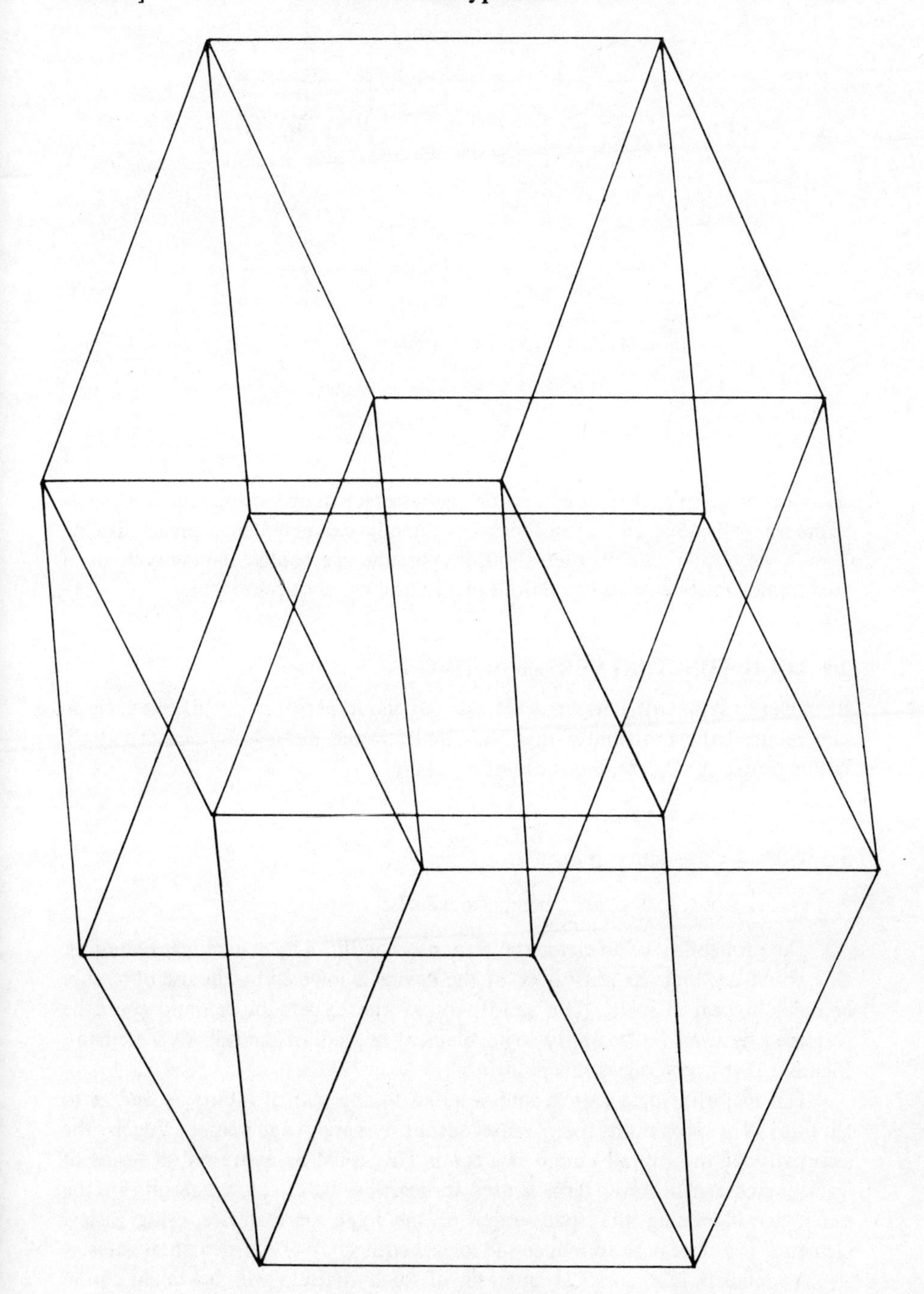

Fig. 2.9 – A view of a tesseract.

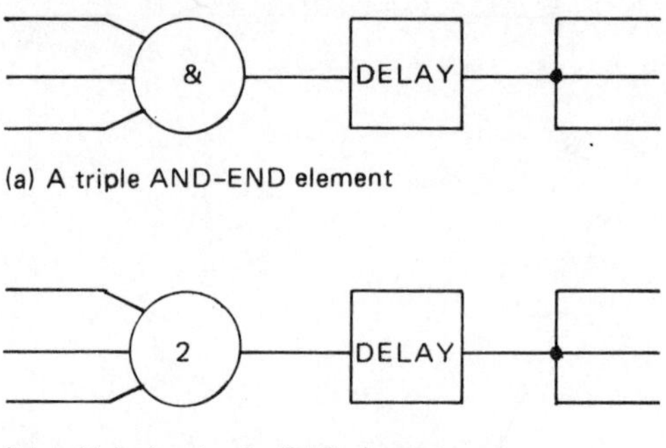

(a) A triple AND-END element

(b) A Majority-Logic-END element

Fig. 2.10 – Triple-input elements.

selection of any planes or cubes within the tesseract can be examined. In fact the entire set of cubes and planes can be directly examined in a plane display. Should we bother to construct a pulsohypercube, we could examine systems of communications between the various planes of a hyperspace system.

2.9 LOGIC FUNCTION VERSUS ACTIVITY

In moving to the triple-input AND gate for the hyperspace circuit, we have in fact reduced the probability that the circuit would display any activity at all. With a simple AND gate, the states of input are:

none, a, b, a&b

but with the triple-input device:

none, a, b, c, a&b, b&c, a&c, a&b&c

The probability of an elemental response is reduced by a significant amount. One could say that the sensitivity of the device is lowered by the use of a more complex logical element. The sensitivity of the hypercubic circuit could be increased by use of a 'majority logic' element in place of the full AND element. Figure 2.10(b) symbolizes this point.

The majority logic gate would respond to any pair of inputs as well as to all three. The probability of a pulse output is therefore increased, that is, the sensitivity of the overall circuit is greater. This could be expressed in terms of 'gain' increase. Whichever term is used to describe the effect, the results are the same: by increasing the space-order of the hyperspace device using logical elements, it is necessary to reduce the logical stringency of the elemental gates.

A detailed mathematical analysis of such matters will be taken up in Chapter 3.

2.10 MULTIDIMENSIONAL ELECTRONIC CIRCUITRY

There are many examples of electrical and electronic circuitry in which high-order space models of the networks are used in both design and development. Particular examples arise in selection circuitry such as in telephone exchanges and in computer memory. It cannot be too clearly emphasized that the actual devices are constructed of normal three-space materials existing in time – the familiar world in which we live.

A parallel may be drawn in classification systems such as in libraries and museums. Consider a group of rock specimens; there are numerous characteristics which associate or separate samples. Rocks may be porous, amorphous, crystalline, agglomerate, grainy, colourful, and so on. Some of these characteristics can be regularly or systematically graded on agreed scales. Others like colour have no such clear-cut grouping. However, each characteristic may be treated mathematically as a dimension in some space model, one new orthogonal space dimension for each additional characteristic. The objects are 'real', that is, normal three-space; it is our descriptions of them that are represented as high-order space systems. Use of the multispace models helps us simplify our view of many a complex system and, especially in electronics, aids in the design process.

In the days of 'discrete electronics', design of the 'core' memory of a computer presented many interesting problems and a remarkably clear understanding of them could be gained by considering the core store as a model having seven orthogonal spacial coordinates. The reasoning is as follows:

For reasons of electrical limitations, it was convenient to have core store organized in 4096-word blocks. For an n-bit word of storage there would be 'n' planes of magnetic beads. The same X–Y position of every plane must be addressed to select a 'word'. In-plane selection was by 'rows' and 'columns' – the X and Y values being coordinates within the plane.

To minimize the number of electrical connections between the selection control circuitry and the 'stack', each edge of a plane was addressed by a pair of x–y coordinates. Thus the x–y coordinates along the X axis were most simply viewed as a normal orthogonal pair in space forming a minor plane normal to the X axis. Similarly, a second plane provided for selection on the Y axis. The Yx,y plane is quite independent of the Xx,y plane and hence these two planes cross nowhere within the selection space. The X axis possesses two orthogonal selection vectors, as does the Y axis – a total of six spacial dimensions. The memory plane is able to provide one selected bit. The set of planes provides a word – orthogonal to the X–Y selection axes. This gives the total of seven space coordinates. Analysis of the core stack in this way enables us to isolate any two or three (or more) selection vectors. This was particularly important with the magnetic core stack whose lines required bi-directional current drive.

Multidimensional analysis is a well-established feature of electronic design procedures. In the realm of digital electronics we use logical connectives and spacial relationships to enable fabrication of complex systems in which logic holds the key to operation. Systems fail in response to a component failure. We avert certain classes of failure by the use of more complex logic for error detection and error correction.

In post-digital electronics, we shall not only use multidimensional analysis as a design aid, we shall often design systems to operate in high-order spacial modes. We shall move further from conventional logical systems so as to reduce the dependence of systems on component failures.

However, for the moment, we shall consider certain forms of sensor which will find application in later work and shall take a look at what form a pulsatory matrix processor may take.

2.11 PULSOR SENSORS

A considerable amount of work has been conducted on optical sensors which produce 'standard pulses' at rates dependent upon the illumination of the photosensitive element of the circuit.

Arrays of 'pulsensor' elements have been constructed using miniature photocells and work on these goes back to the early 1960s. They are described in the early patent of the pulsor system.

The details of the fundamental pulsensor circuits displayed essentially simple forms. There are very sound technical reasons for using such simplicity despite the apparent drawbacks. One of the principal considerations is reliability, another is related to achievable packing density, and another is that precision is not ultimately a requirement for the individual circuits of the systems we shall study later in this book. The non-linearity of the most elementary circuits for pulsensors is often a distinct advantage – and something which one would have striven to design into a sensor.

So, for the present, if you can accept that the circuit of Fig. 2.11 *could* do something – however badly by modern equipment performance standards, we can actually apply it and analyse the behaviour of a number of rather complex systems.

2.12 MULTIPLANAR PULSATOR SENSOR-ACTIVATOR

Consider a plane containing regularly-spaced pulsor elements. Let there be a simple sequence of connections from the elements of the one plane to another and so-on to fill a cube with such elements. The concept is illustrated in Fig. 2.12.

The form of interconnections between an element of one plane and the next plane is that an element is connected to some adjacent elements in the succeeding plane. Given that there are sufficient interconnections to support

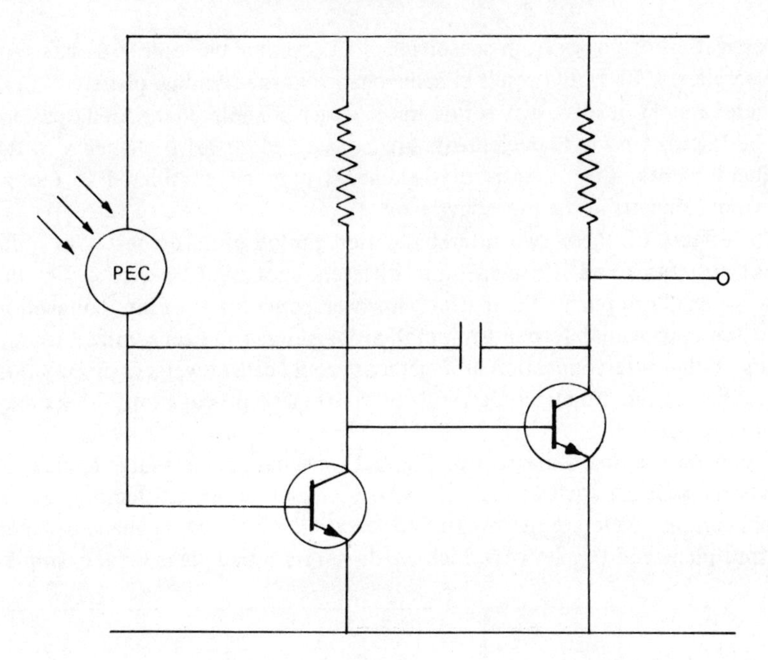

Fig. 2.11 – Photo pulsensor – rudimentary circuit.

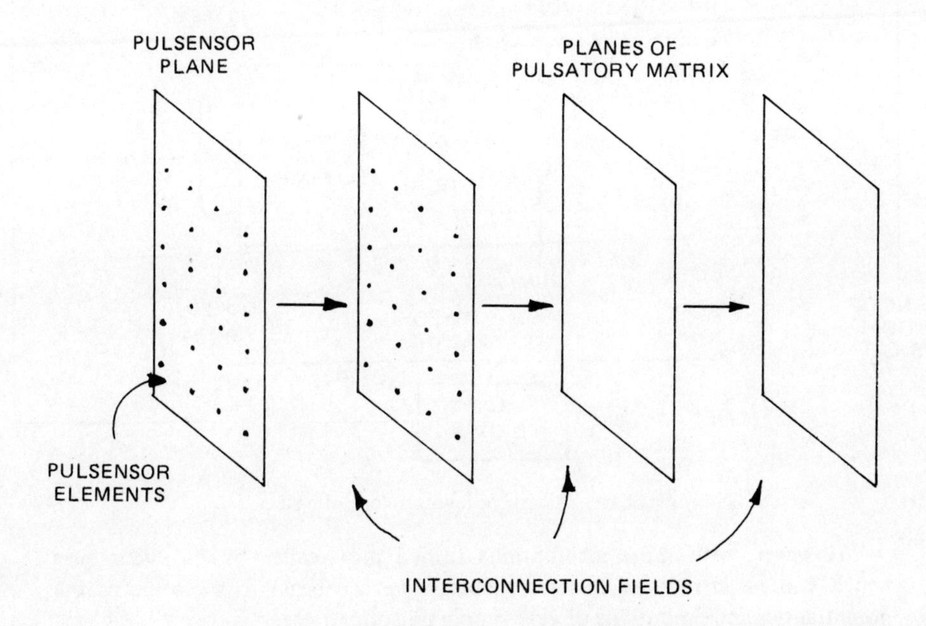

Fig. 2.12 – An elementary pulsensor pulsatory matrix.

the propagation of pulses from sensor plane to terminal plane, then a change of the sensorplane state would result in some change at the terminal plane.

There are, in reality two forms upon which an interconnections pattern could be based. One is that elements are considered to drive elements in the subsequent planes; the other is that elements may be considered to receive pulses from elements in the preceding plane.

The effects of these two interconnection philosophies do result in quite different structures and hence in quite different modes of operation. For the present we shall not pursue the matter. However, some forms of interconnection could cause approximately point-to-point propagation from sensorplane to output plane. Other interconnection arrangements could cause divergent propagation through the matrix, others could result in convergent propagation, others may cause skew propagation.

In the oversimplified diagram of Fig. 2.13, we have made a large number of assumptions such as: we can make it work; we can define satisfactory sets of interconnections; there are meaningful connections which can be made between the output plane and the devices which are driven from that plane – for examples.

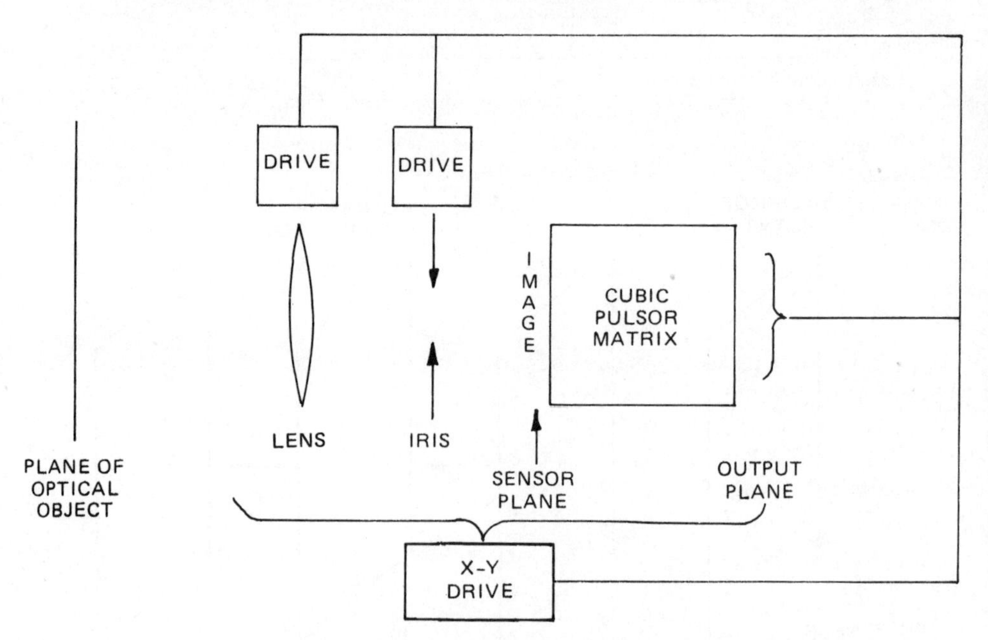

Fig. 2.13 – A basic optosensor control system.

However, with those assumptions turned into reality by the techniques which will be introduced in later chapters, we can begin to see some of the potentialities and limitations of very simple pulsormatrices.

The sensor plane should present no real difficulty in concept. The nature of a standard form of camera optics is straightforward. The idea of driving each of the elementary optical parameters presents no difficulty. Therefore, if the matrix will actually propagate signals of some sort, we should at least be able to control the aperture of the iris by a direct feedback control system. The normal control stability and other performance criteria must of course be evaluated and built into the system. Having achieved that step, the next to consider would be the control of focus. There is little more to focus control than to adaptation to ambient level control.

Assume that the field of view of the system contains a simple, normal plane of high-contrast information such as simple lines – maybe a centralized crossed pair of lines. The effect of focussing an optical device will be to produce a high-contrast image at the sensor plane. Analysis of such systems generally discloses that some space function of the image maximizes at the in-focus condition. A number of such devices have been in use in ordinary cameras for some years.

It is very proper to enquire into the performance of the control system (if such it is) of Figs. 2.13 to varying stimuli (inputs) and to the presence of noise in the system.

No order, class or degree of servo control is indicated in the oversimplified diagram. Such a device was constructed during the early 1960s and could be made to operate crudely to simple, slow demands. For satisfactory operation, the pulsor matrix would need to derive the necessary terms for servo control and stabilization, and to apply these with suitable temporal response, organizing these various signals to suit the drivers of the four actuators.

As to the effects of noise, there is nothing to suggest that the system would remain stable in the presence of noise. Rather more intricate circuitry is required. Discussion of such matters must be delayed for consideration in later chapters.

If then, we are able to reduce a focus-dependent signal component from the matrix, it is quite reasonable to suppose that this parameter of the optical system can come under control.

By further theoretical reasoning along the same lines, we can, and shall, develop techniques for reducing from the matrix processor, signals designed to maximize at image centralization. From such signals, it should become possible to devise X–Y drive to cause image centralization.

There is nothing novel in these concepts at the present state of electrotechnology. Control systems of all these types are known and are in regular use but with the control signal reduction being by either analog or digital computation. In fact, a pulsatory optocontrol system of just the type under discussion was set up by the Author in the early 1960s. The cost of fabrication and the coarseness of the resolution attainable at that time precluded the adoption of the pulsor technique. Interest is reawakened now that a number of suitable microfabricated devices are available. The production of advanced forms of pulsormatrices appears quite feasible.

However, before we can proceed to analyse circuits and devise interconnection plans, there is a certain amount of theoretical work to be covered. In Chapter 3, we shall derive design criteria for the basic elements. The AND-END element is unsatisfactory in a number of respects and improved forms must be derived.

2.13 THE POSITIVE PRINCIPAL IN ACTION

Each of the circuit elements described in this chapter can fail. Failures would occur within the interconnection fields were such matrices to be constructed. What would be the effect of elemental failure on the operational characteristics of the systems?

In each case, the effect could be traumatic. However, very minor alterations to the electrical conditions under which the elements were operated could so alter their behaviour as to make them vastly more active. Then noise and other effects could (and in the early experimental work, did) ensure pattern circulation even after input streams were terminated. Such 'supersensitive' (unstable?) operation may, in a sufficiently complex circuit, render the overall device less liable to fail on failure of a single element. But this is very loose talk. Our designs must be stable; therefore all the systems envisaged in this chapter are subject to failure on failure of any component.

3

From Logic to Probability

3.1 APPROACHES TO ANALYSIS

The analysis of a multiplicity of logical circuits operating asynchronously is hampered by a lack of information on the times of arrival of individual pulses. Apart from the well-understood forms of logical analysis based upon combinatorial Boolean logic, the most common techniques involve frequency-domain analysis with its varieties of transforms and operations.

In early attempts to analyse the behaviour of the original pulsorcube, the use of pulse-frequency maps was considered. Again the sheer complexity of the pulsorcube precluded adoption of such techniques.

Because we have so little information to go on and the complexity of the combinatorial relationships is so high, it is desirable to find an alternative strategy. Rather than consider pulse-frequency (actually meaningless in an asynchronous system) or pulse-rate, consider the probability that a pulse is active at a given point in time. The mathematical techniques are then the application of combinatorial probabilities. This is rather less formidable than other methods of description.

It cannot be too strongly emphasized that we use mathematical *descriptions* of our world and manipulate these descriptions in an attempt to understand existing phenomena and form bases for design and control. The famous Fourier Transform techniques are often misapplied by assuming that the transform IS the system and not just a description.

3.2 LOGIC – TO – PROBABILITY TRANSFORMATION

3.2.1 Logical Symbolism

Our application of Boolean Logic will be of course to electronic switching circuits after Shannon. However, the techniques are those of George Buffery who paved the way for the development of digital computers by his re-analysis of the Boolean system. Strictly then, our convention will be those of 'Buffery Logic'.

We require a symbolism for Logical Constants, Variables and Operators. Using these symbols, we can express the concepts of pulsatory matrices and networks and derive a set of transforms by which the work of analysis and design can be undertaken.

3.2.2 Logical Constants

There are two:

TRUE which implies the presence of a pulse at a given point in time (a pulse has a fixed duration in pulsormatrices).

FALSE which implies that no pulse is present at a given point in time.

3.2.3 Logical Variables

The pulsor devices have a number of inputs and a number (usually one) of outputs. We thus deal with pulses (events) on signal lines. A signal line carries events to device inputs, and we shall use the first few capital letters of the alphabet to denote such events. Thus 'A', 'B', 'C' etc. will represent the states of the various inputs. Output will be symbolized by 'O', or more commonly 'o', in actual examples.

Thus A may be TRUE (pulse present) or FALSE (pulse absent) at a given point in time.

3.2.4 Logical Operators

The following conventions will be used:

Either A or B or both A and B (TRUE) will be A v B

This is the **Logical inclusive OR**

Both A and B (TRUE) will be A & B

This is the **Logical AND**

Logical expressions formed from the above operators and variables are dealt with in the usual manner such as:

A triple-input device will provide an output 0 when:

(A v B v C) is TRUE

also when (A & B) v (A & C) v (B & C) v (A & B & C) is TRUE

However, our actual pulsatory elements provide an output only on termination of a TRUE condition. Using the symbol 'd' to imply 'one pulse time delay', terms such as

A & dB actually hold O FALSE during time *T*b.

We shall therefore, retain a simple approach to circuitry and analysis, using conventional switching circuit theory as a guide to the main theme of our work.

Other useful operators will be symbolized as:

A variable (or expression) inverted will be $\bar{E}$
This is the **logical NOT**

Either A or B but not both will be $A \underline{v} B$
This is the **logical exclusive OR (XOR)**

A signal condition inhibiting output will be $\underline{E}$

The difference between $\bar{E}$ and $\underline{E}$ is that the one would provide a TRUE output in response to a FALSE input whereas the other simply prevents the appearance of a TRUE ouput during existence of the inhibitory condition.

3.2.5 Pulse Rate, Frequency and Probability

The frequency of pulse occurrence may be expressed in terms of n pulses per second. When our pulses are of standard duration, the pulse rate or frequency may be expressed as '1-in-n' pulse periods (or time-slots). This is a 'normalized' way of expressing the rate of arrival of pulses.

Using the latter form, we can relate normalized frequency to the Probability that a pulse is present at a given point in time.

Thus, for a frequency of 1 in n pulse periods
the probability of a pulse is 1 in $n = 1/n$

Probability of course always lies in the range 0 to 1.

With pulsatory devices, the probability of a pulse can never exceed 0.5 and is usually very much less than this. The reason is that no element can deliver an output in two successive time-slots.

The probability P of an event E will be expressed as P(E). This will carry a numerical value in the range 0 to 1 expressed as Pe.

Thus the probability of event P(E) is Pe.

The range of probability values to be found in our work will normally be in the range 10^{-4} to 0.1 though values outside this range will not be uncommon.

3.3 COMBINATORIAL LOGIC AND PROBABILITY TRANSFORMS

Our circuitry is composed of multiple-input devices, each input line of which carries pulses of a given (non-synchronous) frequency or probability. For two-input elements the two main types for initial consideration are the elementary AND and, because of the nature of certain circuitry to be used, the 'exclusive OR' element (symbol $\underline{v}$). A number of other elemental types will also receive attention:

1. Output produced as a result of a signal on either (but not both) line. This is the XOR element. The probability of an output is:

$$P(\mathrm{A}\ \underline{\mathrm{v}}\ \mathrm{B}) \text{ has the value } P\mathrm{a} + P\mathrm{b}$$

Functionally: $$P(\mathrm{A}\ \underline{\mathrm{v}}\ \mathrm{B}) \text{ transforms to } P\mathrm{a} + P\mathrm{b}$$

From this it is seen that the XOR function transforms to the sum of the probabilities or:

$$\underline{\mathrm{v}} \text{ transforms to } +$$

2. Output dependent on both inputs (logical AND):

$$P(\mathrm{A}\ \&\ \mathrm{B}) \text{ transforms to } P\mathrm{a} \times P\mathrm{b}$$

or: $$\& \text{ transforms to } \times$$

The associativity and communitivity laws hold in both logical and probabilistic domains.

3. Outputs produced according to the inverse of a signal. This is the logical NOT:

$$P(\bar{\mathrm{A}}) \text{ has the value } 1 - P\mathrm{a}$$

4. Outputs are inhibited by the signal $\underline{\mathrm{A}}$:

$$P(\underline{\mathrm{A}}) \text{ has the value } -P\mathrm{a}$$

Table 3.1 is a collection of certain interesting logic to probability transforms which will be used in the assessment of the performance of a number of circuits as pulse elements.

Table 3.1

Logic to probability transformations.

Function		Transform			
1. *Buffery forms*					
XOR	$P(\mathrm{A}\ \underline{\mathrm{v}}\ \mathrm{B})$		$P\mathrm{a} \times P\mathrm{b}$		
AND	$P(\mathrm{A}\ \&\ \mathrm{B})$			$P\mathrm{a} \times P\mathrm{b}$	
NOT	$P(\bar{\mathrm{A}})$	$1 - P\mathrm{a}$			
NAND	$P(\overline{\mathrm{A}\ \&\ \mathrm{B}})$	$1 -$		$P\mathrm{a} \times P\mathrm{b}$	
OR	$P(\mathrm{A}\ \mathrm{v}\ \mathrm{B})$		$P\mathrm{a} + P\mathrm{b}$		$-P\mathrm{a}^2 P\mathrm{b} - P\mathrm{a} P\mathrm{b}^2$
2. *Derived forms*					
AND-END			$P\mathrm{a}$	$P\mathrm{a} \times P\mathrm{b}$ Limit 0.5	
Asynch OR-END			$P\mathrm{a} + P\mathrm{b}$		$-2P\mathrm{a}P\mathrm{b}$

3.4 FURTHER NOTES ON LOGICAL ANALYSIS – DeMORGAN'S THEOREM

The Theory of DeMorgan is commonly used to permit the transformation of one logical operator into another – notably the operators AND and OR which are related by the following:

$$A \,\&\, B = \overline{\bar{A} \vee \bar{B}}$$

and

$$A \vee B = \overline{\bar{A} \,\&\, \bar{B}}$$

Whilst this is an extremely powerful tool in the design and development of electronic switching circuits (logic circuits), its use in our asynchronous regime is not necessarily valid. The reason is simply that the classical theorems relate to 'static' conditions and take no account of for instance transition times, percentage overlap to satisfy a conjunction and so on.

Again, in our system there is no condition which provides a TRUE output for longer than a 'standard pulse time'. It would be rather more accurate to use the concept of 'BUT NOT' rather than the classical 'AND NOT'. The result is that we *subtract* the probability of certain conditions rather than multiplying by $(1 - Pc)$.

3.5 PROBABILISTIC TRANSFER FUNCTIONS

From the foregoing transformations we can derive a **Probability Transfer Function** (G) such that the ratio of the combined (output) probabilities to the individual (input) probabilities is expressed as:

$$G = P_o :: P_i$$

Quite obviously, for an n-input device:

$$P_o = C_1^n(P_i)$$

where C is the combinatorial of all such terms as i for $1 = 1,n$.

The Probability Transfer Function (G) may be taken to be the 'gain' of the system and can be applied as such to both elements and to assemblges of elements.

Of interest in analysis is the elemental condition which provides unity gain $G = 1$. This will be found to be both hardware dependent and to vary with the incoming signals. Being a significant factor in design work, we shall adopt a symbol for it:

$$G = 1 :: Po/Pi = 1 :: P1.$$

3.6 PROBABILISTIC EQUATIONS

The equations derived in the foregoing may be applied to the special cases of the asynchronous logic elements (AND-END and OR-END) with rather interesting results.

The probability of a pulse occurring at a given point in time is related to the pulse frequency. If a pulse were to occur every ten pulse periods then the probability is simply 1-in-10. With a two-input AND-END circuit operated synchronously, then the output pulse probability woul be $P_a \times P_b$, that is, 1/100. From this fact, that is, that the elemental gain is 1/10, it is obvious that the synchronous pulsorcube is inactive.

For the asynchronous AND-END element however, the probability of an output is increased because there is a reduced overlap requirement. Taking r as the overlap required between pulses on the two inputs, the equations are modified as follows:

$$P(\text{A} \,\&\, \text{B}) = (2 - r) \times (P_a \times P_b)$$

When the overlap is unity as in the synchronous application, the original result applies.

The element will have a gain of unity when P_i is $0.5 + r$.

We can relax the conditions still further by using OR-END rather than AND-END elements. When this was tried, other stable states were found. At first sight it may be felt that use of the OR-END function would allow a greater variety of operational conditions. However, the OR-END elements require all input pulses to terminate before an output pulse can be delivered. The OR-END function supresses all overlap states, thus operating more like an exclusive-OR (XOR) device. The truth table for the AND element has one out of four possible active states; the XOR has two out of four whilst the OR function has three out of four active states. Used asynchronously, there will be more than a one in four chance that conditions suit an AND-END output whilst there is less than a two in four chance that the OR-END element will deliver a pulse. A little surprisingly then, the type of element selected has less effect on system performance than one may have expected.

3.7 PROBABILISTIC ANALYSES

3.7.1 The Synchronous Two-input AND

$$Po = Pa \times Pb$$

When $Pa = Pb$ $(= Pi)$, we have:

$$Po = Pi^2$$

The Gain may be calculated from:

$$Po/Pi = Pi = G$$

that is, to satisfy the condition that $Po = Pi$ or $PTF = 1$

$$P1 \text{ must } = 1.$$

This is clearly impossible.

A pulsorcube constructed of such elements – or of AND-END elements cannot sustain any form of activity. This lends support to the logical deduction arrived at in Chapter 2.

3.7.2 The Synchronous Two-input OR

$$Po = Pa + Pb$$

When $Pa = Pb \; (= Pi)$ we have:

$$Po = 2 \times Pi \text{ from which:}$$

$$Po/Pi = 2.$$

The unity transfer condition $P1$
The transfer function is 2, that is, linear gain
The gain of this element could support pulsorcube activity.

3.7.3 The Asynchronous Two-input AND-END

Given that this element will produce an output on termination of an input coincidence (of overlap ratio r), we see that:

$$Po = (2-r)(Pa \times Pb)$$

For the condition $Pi = Pa = Pb$

$$Po = (2-r)\,Pi^2$$

whence $\quad Po/Pi^2 = 2-r$

This is a non-linear device so, to determine the unity-gain condition where $Po = Pi = P1$ we see that:

$$P1 = 1/(2-r)$$

For an overlap ratio r of say 10%, $P1 > .5$ which is sufficient to support pulsorcube activity.

3.7.4 The Asynchronous Two-input OR-END

The conditions are similar to those for the synchronous version resulting in a Gain of 2 – sufficient to support pulsorcube activity.

3.7.5 The Asynchronous Three-input Majority-Logic element

The logical equation is:

output occurs when
(A & B) v (A & C) v (B & C) terminates.

Which transforms to the probabilistic form (including overlap):

$$Po = (Pa \times Pb + Pa \times Pc + Pa \times Pc) \times (2-r)$$

For the condition $Pa = Pb = Pc = Pi$ we have:

$$Po = (2-r) \times 3Pi^2 \quad \text{a non-linear device.}$$

The unity-gain point is reached when: $Po = Pi$ and

$$Po/Pi^2 = 6-3r = 1/P1$$

whence $\quad P1 \quad P1 = 6-3r$

For 10% overlap, $P1 = 0.175$ representing a gain of 1. Use of this element in a pulsorhypercube would result in an extremely active system.

3.8 CONCLUSIONS REGARDING LOGIC PROCESSOR ELEMENTS

Both from logical reasoning and latterly via probabilistic analyses, we have seen that logical elements and combinations of various logical elements can be coaxed into operation as matrix processors of certain types. We are able to assess the performance of potential designs of both synchronous and asynchronous systems.

However, even with careful design, the suitability of these devices as fail-proof processors is highly questionable. Designs of considerable processing power would contain a variety of types of element and would hence become extremely difficult to fabricate.

We shall consider the feasibility of designing matrices holding the equivalent of 10^7 or 10^8 processing elements in which even 10% elemental failure rates should not affect overall performance. This could not be achieved using conventional logic circuitry.

The **logic-to-probability transform** technique has provided us with a means for assessing the performance of complex systems. The excursion into the realms of logic fantasy have not been in vain. During the 1960s the author derived immense enjoyment considering and designing multidimensional electronic matrices. A number were constructed, in part for amusement and in part for the experimental confirmation of theoretical ideas. However, for the present we must move on to more profitable investigations.

3.9 THE LEAKY SUMMATOR ELEMENT

3.9.1 The Summator – Probability Gain

The probability transform can be applied to summators as readily as to 'logic' elements. For a simple summator with a terminal count of k pulses, that is, an output pulse is delivered when the total count of input pulses is k, the equations are simply:

$$Po = 1/k$$

Gain is $\quad G = 1/k \quad$ (a distinct loss in fact).

For an n-input summator:

$$G = n/\mathrm{k}$$

which could represent either a gain or a loss.

However, the simple summator is, in reality, yet another form of 'logic element' and will fail all the tests for 'reliability' in networks. Further it is logically a complex device requiring considerable amounts of fabrication space – its packing density in a physical matrix would be rather low.

3.9.2 Leaky Summators

We turn attention then to the analog–digital hybrid – the **leaky summator elements (LSE)**. In this, our 'count' is by the charging of an integrator capacitor by incremental steps determined by the input pulses. The capacitor has associated leakage of its stored charge in an exponentially-decaying mode.

The equation of the simple summator is therefore modified by the exponential decay curve:

$$(1 - \mathrm{e}^{x})$$

where x is determined by both the circuit constants and the rate of arrival of charge increments.

If we are successful in application of this element to our matrix design, the device could be capable of microfabrication in a small volume and should be operable at very high input data rates.

In fact, numbers of these elements have been constructed using 'discrete components' and tested with results to confirm the theoretical analyses – but that was in the early days of microfabrication. With present-day techniques, vast numbers of such elements could readily be assembled in a very small space.

3.9.3 The Leaky Summator Element – Probability Gain

For this element we have: $Po = \dfrac{1}{K}(1 - \mathrm{e}^{-dP}) \sum_{1}^{n} P\mathrm{i}$

where k is the required count of pulses to give an output,
d is the decrement factor,
P is the mean pulse rate expressed as a probability,
n is the number of input lines.

Analysis proceeds as follows. When all Pi are equal then:

$$Po = \frac{n}{k}(1 - \mathrm{e}^{-dPi}) Pi$$

Gain is: $$G = \frac{n}{k}(1 - \mathrm{e}^{-dPi})$$

However, a further modification is required to allow for the natural recovery period following an output pulse. This can simply be incorporated in the factor d which we shall now derive.

In absence of the exponential term, the gain would simply by $\frac{n}{k}$, the ratio of the number of lines carrying input signals to the required pulse count. These are simple design parameters. The existence of the exponential term shows the gain to be dependent on the inputs. By suitable choice of the factor d, a unity-gain condition can be set at any convenient $P\text{i}$ value.

3.10 APPLICATION EXAMPLE – REGULAR FOUR-INPUT LSEs

3.10.1 A Right Pyramidal Matrix of LSEs

Figure 3.1 illustrates the arrangement of a group of LSEs. A plane of 25 photo-sensor LSEs is assumed to be illuminated evenly so that all elements give the same frequency of output pulses. We take this as a set of outputs of equal probability. In a series of analyses the illumination (that is, output probability) will be swept over a range of values to provide the gain locus. The conditions will be altered to determine the behaviour when one element receives a higher level of excitation. The illumination will be swept as before to assess the performance of the device as the 'spot' falls on elements from the centre to the edge of the sensor plane. A final sweep will be made using a 'defocussed spot'.

The photosensor plane provides inputs to a series of other planes of LSEs, each plane containing a smaller number of elements so that a single output line provides a signal whose fluctuations may be plotted as a set of graphs.

The analyses will show the processing capabilities of a small and simple LSE matrix.

From Fig. 3.1 it is seen that each processing LSE of the matrix receives inputs from four elements of the preceding plane. The equation for each element is then:

$$P\text{o} = \frac{4}{k}(1 - \text{e}^{-dP}) \sum_{1}^{4} P\text{i}$$

Because the conditions of the first 'test' involve equal stimulus of all sensor elements and the matrix is a right pyramidal arrangement of elements, the equation can be expressed as:

$$P\text{o} = \frac{16}{k}(1 - \text{e}^{-dP\text{i}}) P\text{i}$$

For a given k and given d there will be a value of $P\text{i}$ resulting in $P\text{o} = P\text{i}$. This is the unity gain condition for which we have adopted the symbol $P1 = P\text{o} = P\text{i}$.

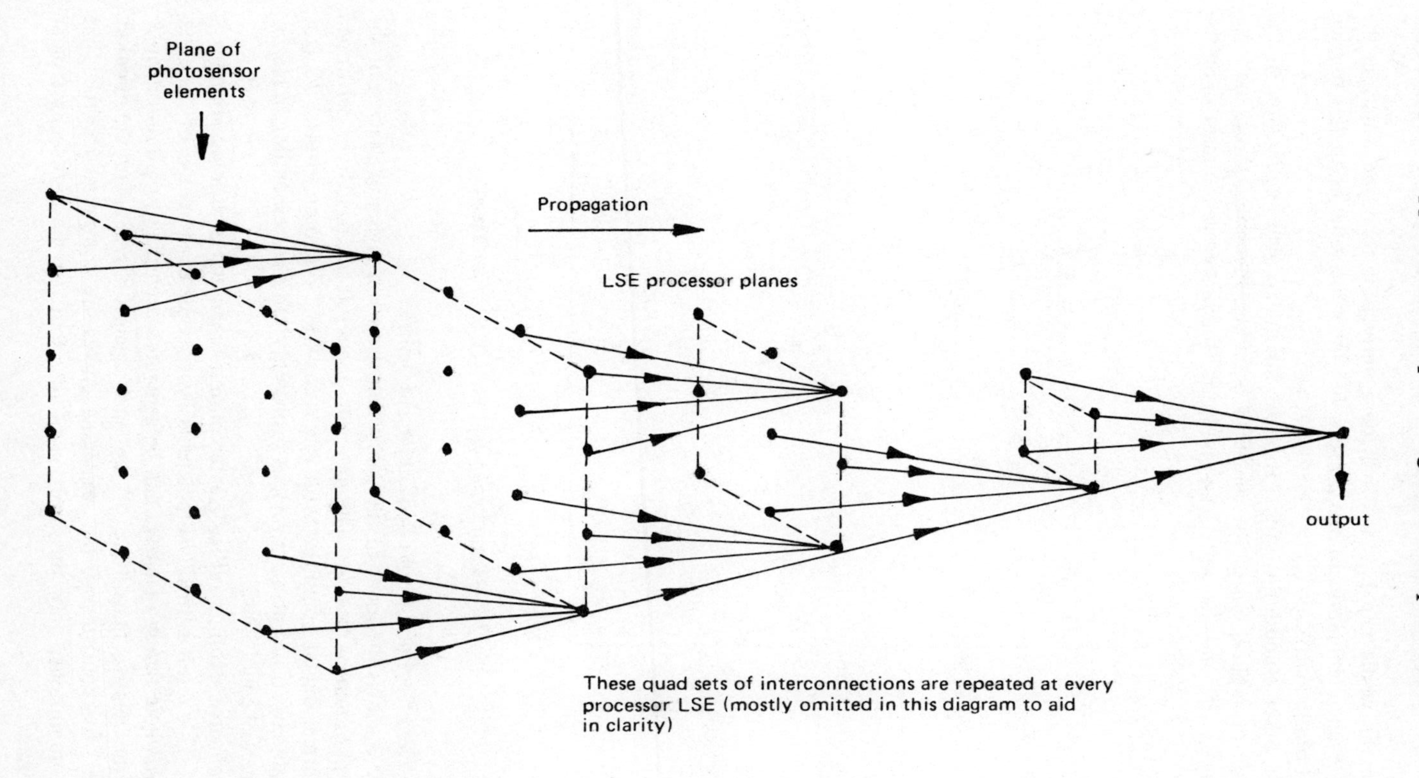

Fig. 3.1 – Right Pyramidal Matrix of LSEs.

For the whole matrix, there will be a unity gain condition which, in the case of even illumination of the sensors, has the same value as applies to each element of the matrix.

For other stimuli the propagation of the matrix will be the sum of the whole set of applications of the elemental equation. Figure 3.2 shows the locus of *Po* against *P*i and the unity gain condition is indicated as a circle (A) on the locus.

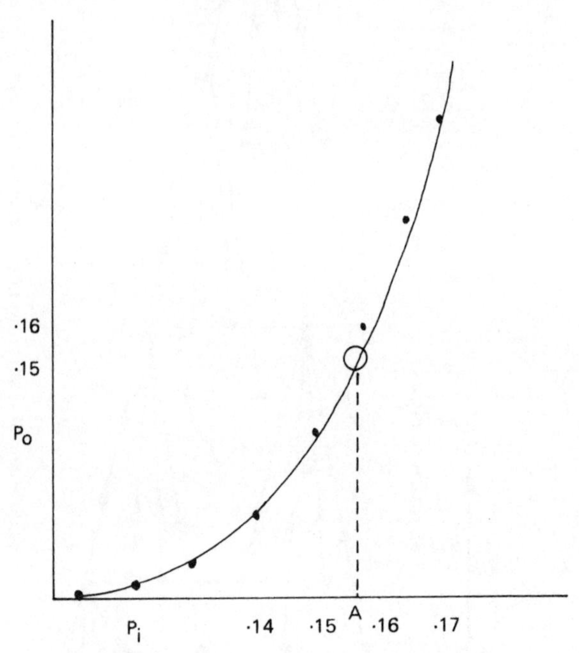

Fig. 3.2 – Right Pyramidal LSE Matrix – Gain Locus.

3.10.2 The Effect of a Bright Spot at Sensor Centre

Figure 3.3 shows the locus of the gain when there is a higher excitation of the central photosensor element. The data were the result of calculations performed on a normal digital computer with values for *k* and *d* chosen to place *P*1 (the unity-gain condition) at a convenient probability value corresponding to reasonable practical conditions.

The curve exhibits rather greater output when the additional stimulus is present. This result is readily predictable by a simple examination of the equations. A whole family of curves can be generated according to alterations in *k*, *d* and the relative magnitude of the 'spot' against the ambient illumination. However, for consistency, the remaining curves are computed for the same general conditions, only the position of the 'spot' on the photosensor plane being altered.

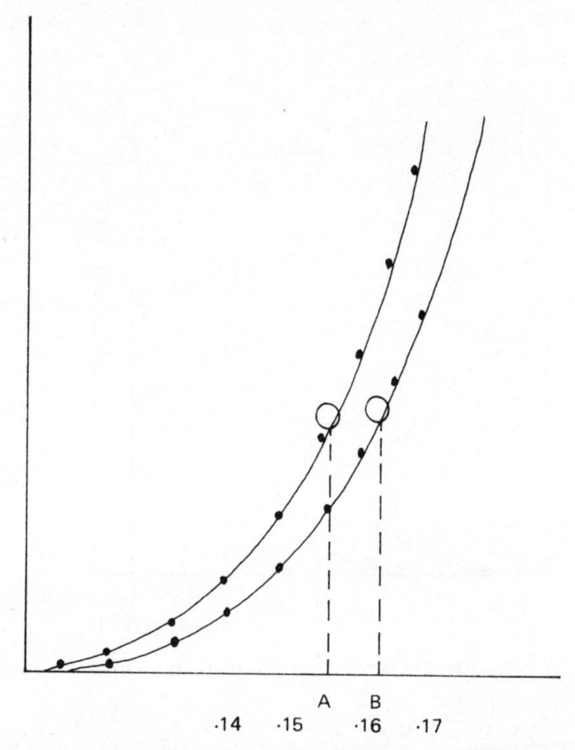

Fig. 3.3 – Right Pyramidal LSE Matrix Gain – Central Spot.

3.10.3 An Off-centre Spot

In Fig. 3.4, the 'spot offcentre' and 'spot at edge' curves are shown. These illustrate that the output from the matrix contains sufficient information to drive both an illumination control device such as an iris, and centering controls as discussed in Chapter 2.

3.10.4 A Defocussed Spot

The final curve of the series (Fig. 3.5) shows the effect of a defocussed spot. The stimulus values of *P*i at the sensor plane used the same total excitation as before – but the excess, that is, spot magnitude, was distributed equally over elements near the centre of the sensor plane.

There is little to observe in the single family of curves which would enable a total control system to distinguish between certain of the excitation conditions. However, a moments reflection will disclose that by driving one actuator at a time from the total signal (*P*o), as each is optimized, the result would be a centrally focussed spot. The device would obey all the normal laws of control systems and a whole set of conventional control circuitry would be required to obtain a rather marginally stable, low-order control system.

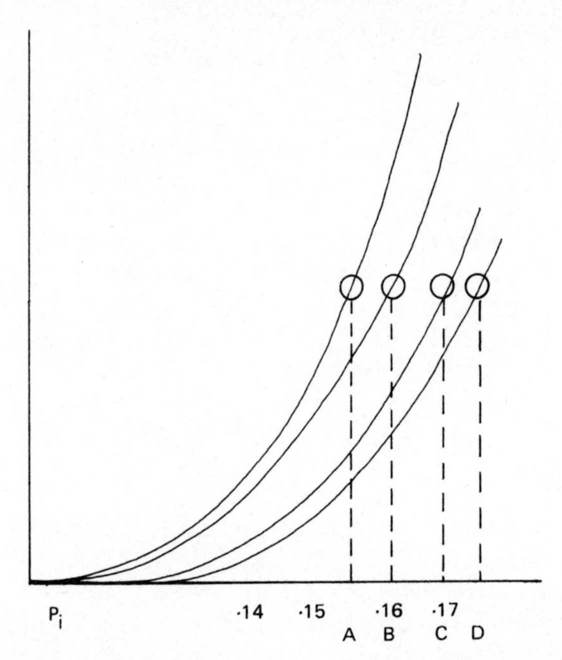

Fig. 3.4 – Right Pyramidal LSE Matrix Gain – Off-centre Spot.

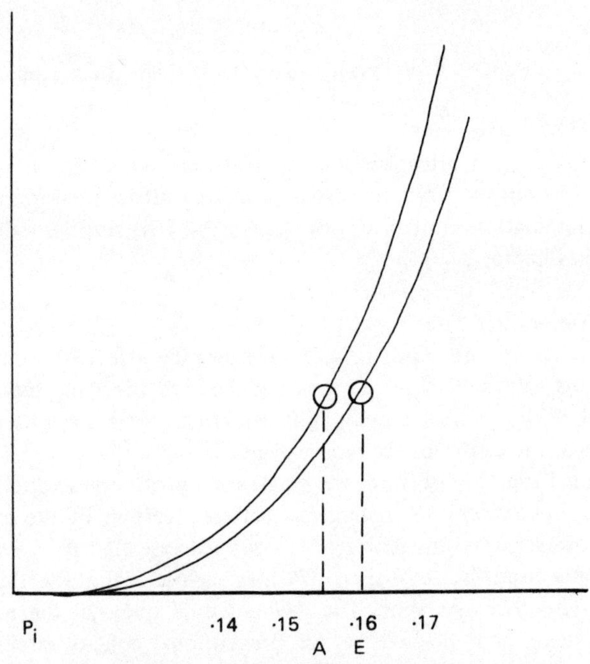

Fig. 3.5 – Right Pyramidal LSE Matrix Gain – Defocussed.

3.11 CHARACTERISTICS OF THE LSE

However, the point has been reached that the LSE does offer a potential for signal processing and can readily handle sets of source stimuli.

The reader may at this point like to conjecture that the right pyramid of LSEs could, under certain conditions, detect moving point stimuli and perform elementary indications of a number of other 'events'.

The LSE gives us a flexible design element with which to construct matrices of pulsatory elements having a variety of characteristics. As will be seen in Chapter 4, such elements are simple in electronic circuitry and hence are capable of high-density microfabrication.

Using the LSE as a matrix design element, we may expect to be able to determine matrices which have an overall gain – the probability amplifier. With this tool we can proceed to design differentiators and integrators which operate either in time or space domains. The result is a set of design bases with which complete systems may be fabricated.

The gain of our elements and matrices is variable intrinsically but we may wish further control. In particular, with an amplifier we must avoid saturation. This can be arranged quite simply by inhibiting output from cells whose inputs exceed a predetermined rate (or probability). Such a control would provide the basis of fully automatic adjustment to ambient conditions and to other external factors.

For the remainder of the book, the LSE will be the major component of matrices. Its inherent simplicity implies reliability – but more imprtantly its adaptability suggests that failure of a cell will have little effect on the performance of a matrix. A curious result of such thinking is that, if defective elements occur in a fabricated matrix, we may not even know. Hence fabricational constraints are lessened and the yield of microelectronic devices will be very high.

4

Elemental pulsator devices

4.1 FOOTNOTE TO ALL CIRCUITS IN THIS CHAPTER

None of the circuits given includes any form of compensation, stabilization of parameters, or even calibration. This is entirely intentional. We are searching for minimum-component-count elemental units capable of mass-fabrication with very high tolerance to variations in component values.

All circuits must operate with low-precision components. There can be no form of calibration 'in-service' under the conditions of pulsatory operation. Use will be non-synchronous. Only power-rail potentials can be expected to be within say 10% of quoted values – and we would expect a completed system to perform well with power levels varying over very considerable ranges.

The traditional approach to the design of ICs (provision of noise-immunity and close tolerance in all parameters) becomes quite unsuitable for pulsatory system design. We would expect a typical 'pulsatory IC' to be large, with an area measured in square inches. In manufacture we would actually prefer that at least 10% of elements were defective. A silicon slice would not be 'diced' as with conventional microelectronics work. We would work with slices, and the manufacturing yield would be exceptionally high due to the tolerance of our systems to defective components.

4.2 SOME ELEMENTARY LOGIC CIRCUITS

Logic elements are fabricated in vast quantities in the microelectronics medium of silicon. Even so, an ever-present design problem is the quantity of 'components' per 'cell'. An example of the problem is to be seen in the development of ROM (read-only memory) and RAM (random-access memory) 'chips'. ROM cells can be fabricated with one transistor per memory cell. With 'dynamic RAM' one transistor suffices, but with the rather more flexible 'static RAM', most designs require three or four transistors per cell. As a result, most microcomputer systems built up to 1981 utilize dynamic RAM with its attendant disadvantages of the need to 'refresh' the contents of the memory frequently.

The sheer economics of design, development, and production has dictated that for low-cost computers, dynamic RAM be used whilst for small-quantity production and 'one-offs' the more obvious choice was static RAM. With the inevitable improvements in fabricational techniques, the relative cost of the two types of RAM is narrowed.

Logic circuits may be constructed using any of a variety of techniques including:

(a) Diode logic
(b) Resistor-Transistor logic (RTL)
(c) Diode-Transistor logic (DTL)
(d) Transistor-transistor logic (TTL)
(e) Emitter-Coupled logic (ECL)
(f) Charge-Coupled devices (CCD)

and, of course, a number of other systems – most having many variants.

Figure 4.1 shows two early forms of logic circuit. The first is based on the use of diodes to form a logic AND circuit, followed by a transistor to restore the loss of voltage levels due to the diodes. The result is AND followed by inversion, that is, a NAND element. The second circuit of Fig. 4.1 was widely used in the early days of computer circuitry as a general-purpose element. It could be made to operate as an AND-element, an OR-element or as a Majority-logic element according to the value of the bias provided.

Practical logic circuits tend to include rather large numbers of components to achieve high switching speeds and noise immunity.

4.3 An RTL CIRCUIT

As this is an investigation into types of suitable circuitry for a new form of electronic device, we shall take a little time to pause over the outmoded RTL system.

Consider the circuit of Fig. 4.2. It contains a transistor which is switched by the three inputs according to the ratio of the resistors and the number of inputs applied. Given that the supply voltage is equal in magnitude but of opposite sign to the bias voltage, the operation of the circuit is dependent upon the ratios of the input and bias resistors. We will assume all input resistors to be of equal Ohmic value. We shall also assume that the transistor has high gain and a high switching speed. To a good approximation then, the value of the bias resistor Rb determines the type of the logic operation according to the following list:

(a) Rb = R/2 AND-gate with NOT, that is NAND
(b) Rb = 2×R OR-gate with NOT, that is NOR
(c) Rb = R Majority-logic with NOT

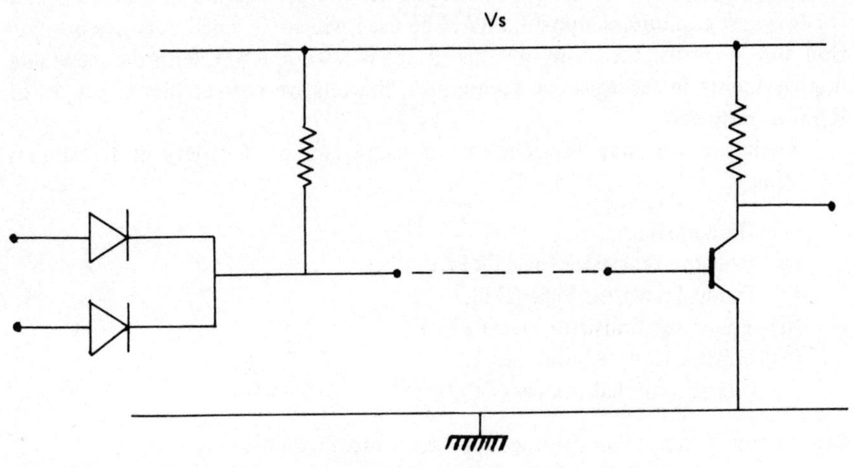

Diode–Resistor AND + Invertor = NAND

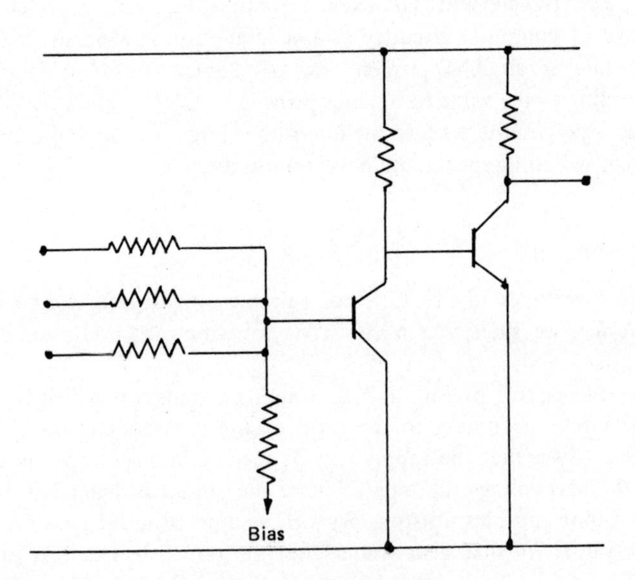

Resistor–Transistor Logic
AND, OR Majority Logic determined by Bias

Fig. 4.1 – Some early logic circuits.

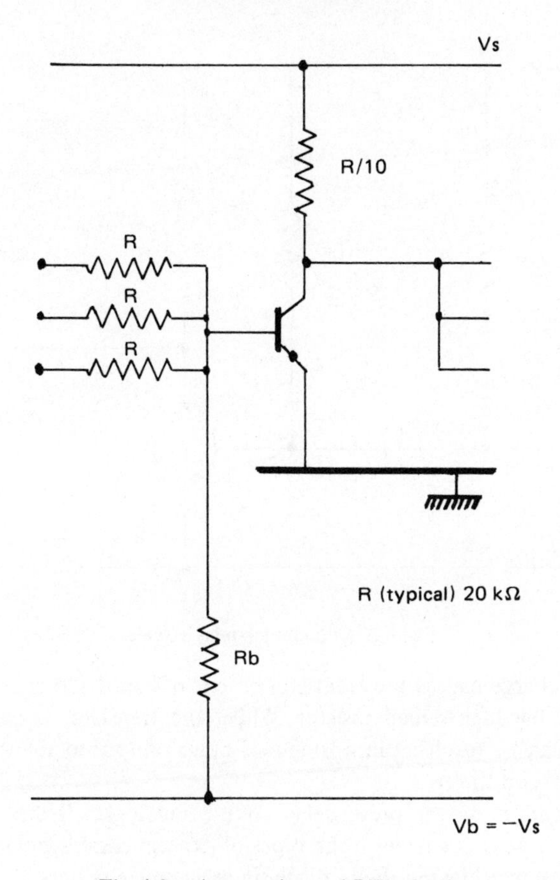

Fig. 4.2 – An experimental RTL circuit.

In passing we note that condition (c) above corresponds with the requirements for 'NOT Carry' in an Adder Circuit.

Because of its simplistic flexibility, an RTL form was selected in early experimental work with the pulsorcube and a number of other pulsor applications. An effect, similar to that of altering the value of the bias resistor, could be achieved using a fixed-value for the bias resistors of a group of elements and altering the bias voltage supply. This enabled a range of experiments to be conducted using a common set of elements.

4.4 PULSE DELAY ELEMENTS

A rather fundamental circuit is given in Fig. 4.3. The transistor is normally-conducting. When the input level rises, the capacitor will charge via base current. The transistor remains bottomed. When the input potential returns to common,

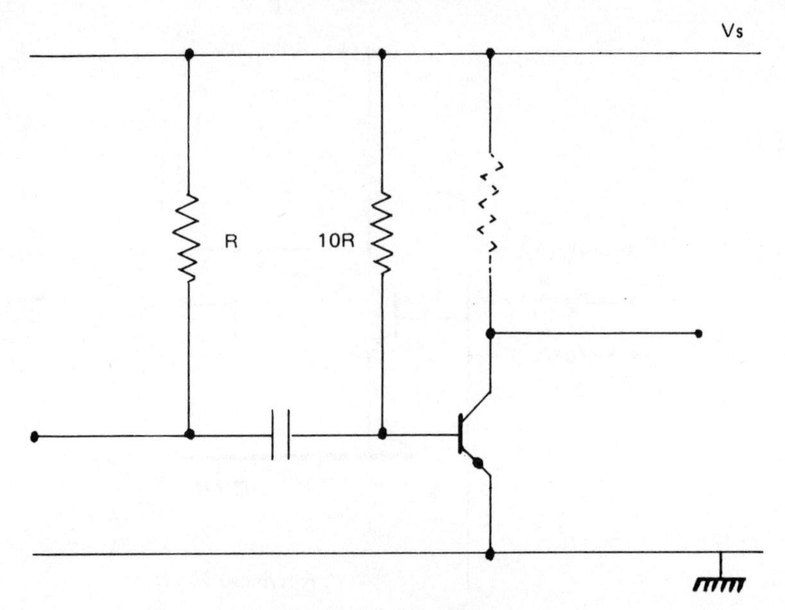

Fig. 4.3 – Elemental delay circuit.

the capacitor charge causes the transistor to cut off unitl the charge has leaked away through the high-valued resistor. Whilst the transistor is cut-off, output potential is 'high' – resulting in a 'standard pulse' output at the termination of the input 'high' condition.

A combination of the pulse delay circuit and logic circuits of the types discussed earlier provides us with the types of experimental logic-delay elements which have been investigated theoretically in preceding chapters.

Much of the early experimental work on pulsor matrices was performed using groups of such circuits. Apart from being compact and using discrete component fabrication, such circuits could well be microfabricated.

The circuit of a basic logic pulsatory circuit element used in early experimental work is given in Fig. 4.4. Circuits of this type were used in the fabrication of a number of the forms of matrix described in Chapter 2. Both PNP and NPN germanium transistors were used in combination with germanium diodes. These components offered very low terminal potential when conducting, thus enabling many variations to be tried in a simple manner. However, the very high temperature drift characteristics prevents the use of such components in a practical device.

Technological advances since the 1960s have provided us with high quality components using silicon as the fabrication base. Many of the difficulties experienced in the early work are thereby minimized. However, there remains much work to be done before we shall see the vast arrays of pulsatory elements required by large systems.

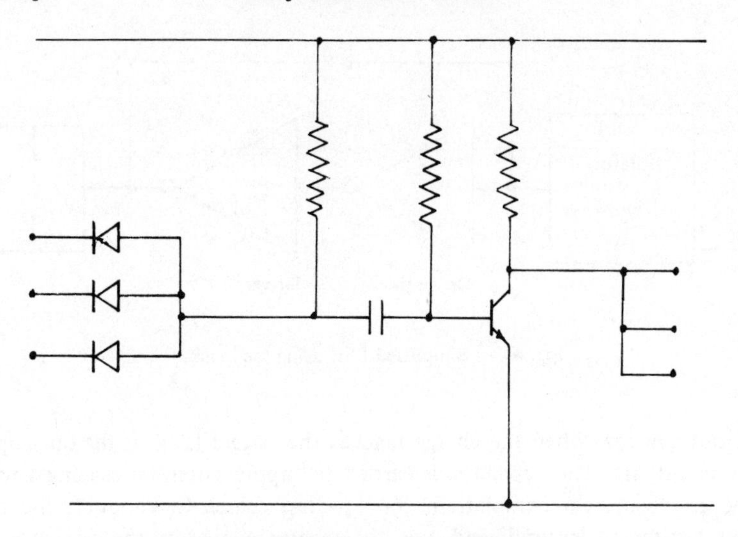

Fig. 4.4 – Basic logic pulsator circuit.

4.5 LEAKY SUMMATOR CIRCUITS

The essential features of the LSE are summarized in Fig. 4.5. Inputs are summed in the capacitor. The trigger element issues an output in response to the summated inputs and discharges the capacitor.

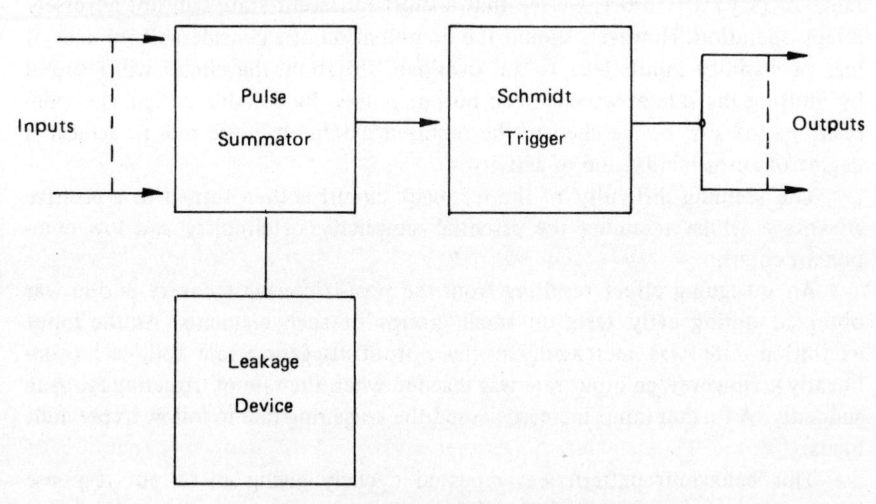

Fig. 4.5 – LSE functional components.

Use of feedback techniques in the design of the circuit can result in considerable simplification as shown in Fig. 4.6. Here, the driver holds one plate of the capacitor effectively grounded so that charge can be accumulated from

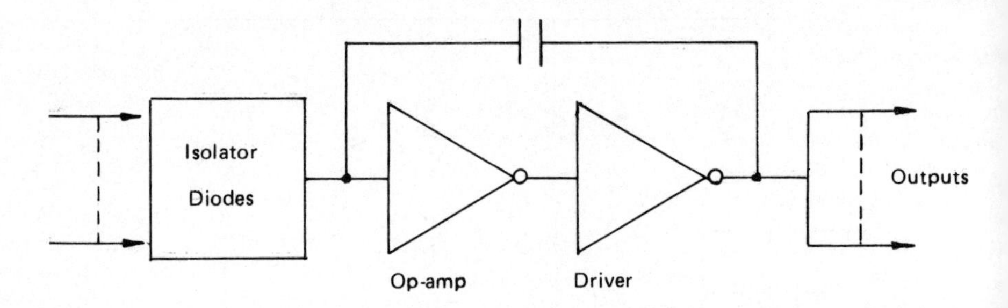

Fig. 4.6 – Simplified LSE using feedback.

the input sources. When the charge reaches the trigger level of the op–amp, the driver is cut off. The capacitor is carried to supply potential causing a reverse charge to occur. On completion, the op–amp ceases to conduct, the driver output returns to 'ground' and now the reverse charge on the capacitor must be removed before 'real' pulse counting can again commence.

4.5.1 Quiescent State

The quiescence of the circuit following a trigger may appear to be disadvantageous until typical pulse rates are considered. Working in the pulse-probability range of (say) .01 to 0.1, we see that a short quiescent state will not adversely affect operation. However, should the combination of a considerable number of high-probability inputs lead to an 'overload' situation, the circuit will respond by limiting the rate at which it can output pulses. By suitable design, the quiescent period can be matched to the required maximum pulse rate to achieve a degree of auto-stabilization of activity.

The seeming difficulty of the feedback circuit is then turned to a positive advantage whilst retaining the essential simplicity of circuitry and low component count.

An intriguing effect resulting from the post-triggering recovery period was observed during early tests on small groups of such elements. As the input excitation rate was increased, so the output triggering rate followed (non-linearly). However, an input rate was reached when the rate of triggering dropped suddenly. A further input increase caused the triggering rate to follow its previous locus.

This behaviour pattern was repeated cyclicly giving an output response having much in common with the aliassing frequencies produced by inadequate sampling of an analog signal. There is, however, a significant difference between the behaviour of digitally-sampled aliasses and the over-excitation of a LSE. Whereas the spectral locus of a simple alias rises and falls as sample rate is reduced, the PRF of the over-excited LSE follows a sawtooth locus.

4.6 LSE FABRICATION

A considerable number of circuit and component arrangements for feedback LSEs has been tested both elementally and in matrices of up to a hundred elements. Essentially similar performance is obtained with all forms tried. The design of such circuits is thus simple and reliable in that individual elements are essentially independent of parameter tolerances in components over a very wide range. The 'count-to-trigger' is readily controllable on one component – the single-input resistor. Numbers of additional inputs may be connected with virtually no change in the fundamental operation of the circuit.

Overall performance of matrices of such circuits can be controlled by a single parameter (say bias) applied to all elements.

The summation charge leakage so essential to the operation of pulsatory matrices and networks is a natural part of the circuit. In the absence of input pulses, the charge can leak away via the input resistor. Thus overall sensitivity and decay are under control of this one parameter.

Figure 4.7 shows that the op-amp of the circuit may be a single transistor or FET of virtually any type. No special type of device is required for the driver, and conventional Integrated Circuits may be used. Transistors may be PNP or NPN, FETs may be P-channel or N-channel, and the diodes have no special constraints.

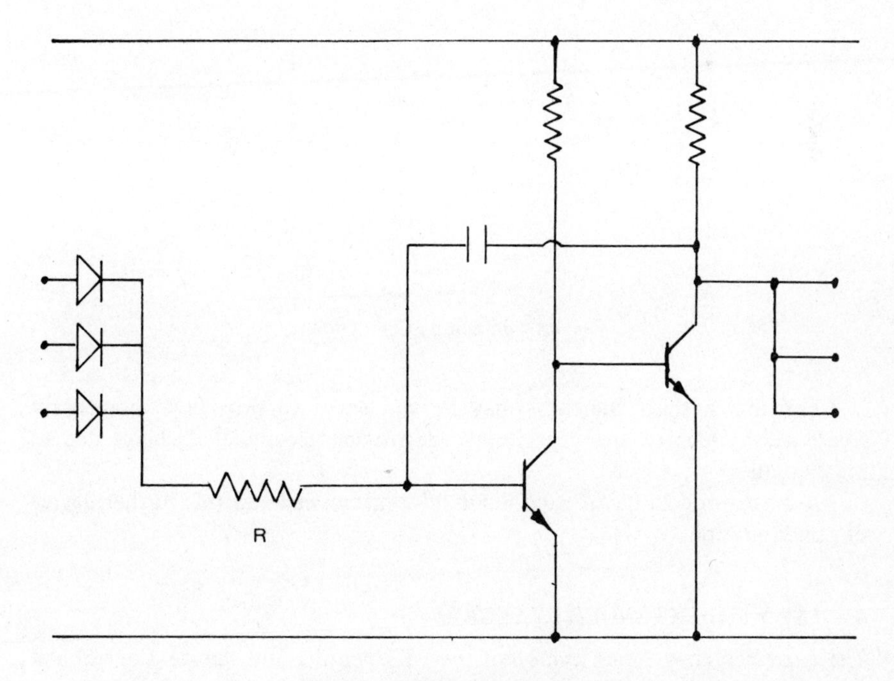

Fig. 4.7 – Feedback LSE circuit.

4.7 INHIBITION

The feedback LSE embodies many of the features required for use as a pulsatory element. However, the inclusion of other attributes may seem desirable. The necessity for auto-stabilization of gain may demand more than just the self-limiting feature provided by post-pulse quiescence. In such cases then, a second form of input may be needed. The purpose of this would be to inhibit operation of the circuit, for example, to induce a quiescent period without the output of a pulse.

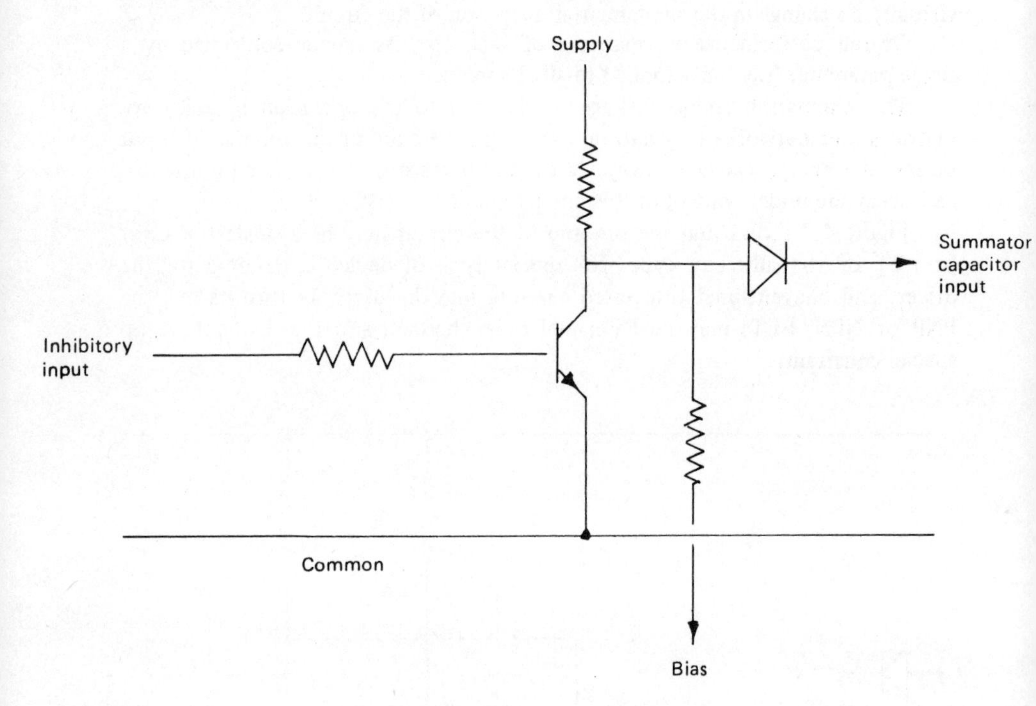

Fig. 4.8 – Inhibitory input circuit.

For this a third transistor may be connected to provide a momentary (pulse-width) bias of the capacitor so destroying accumulated charge due to pulse-count.

A small proportion of such inputs in a matrix can add this further degree of gain limitation.

4.8 LSE WITH EXTENDED FEATURES

The extremely low input leakage of the FET permits both greater control and separation of various LSE parameters and extension of the LSE characteristics.

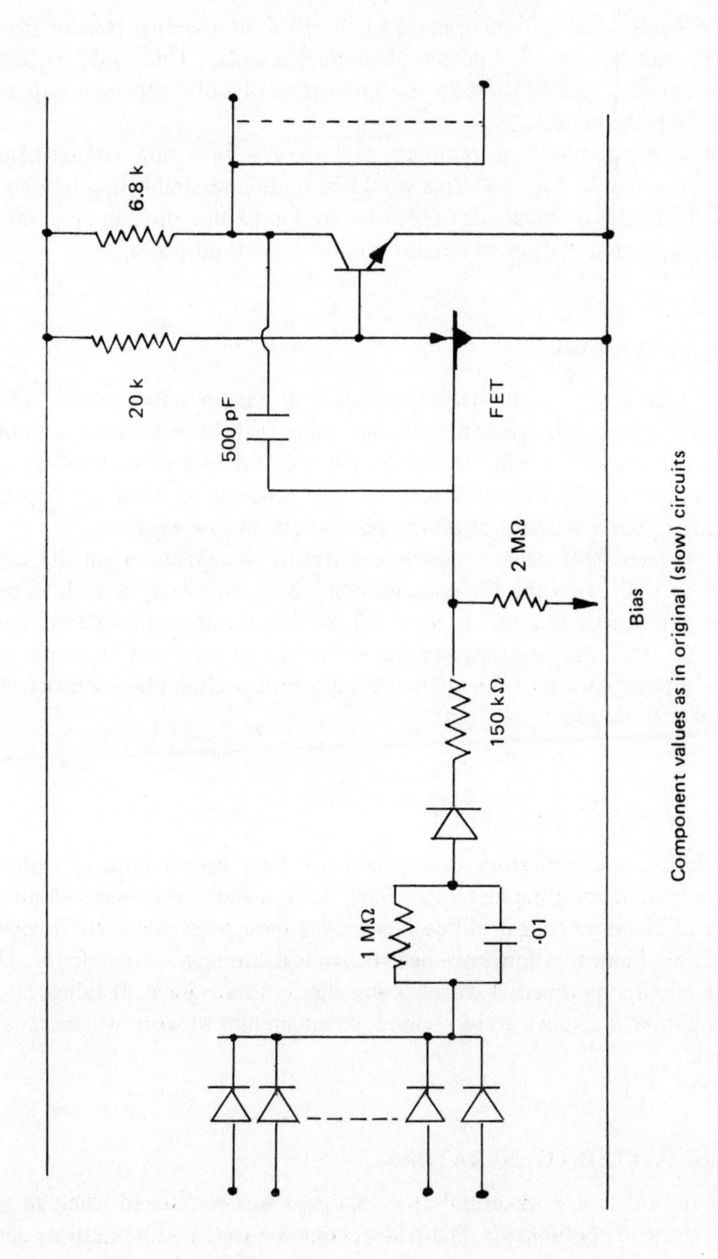

Fig. 4.9 – Early LSE circuit (circa 1964).

There being little natural leakage in the FET, an external resistor may be used to provide a controlled degree of summator decay. This could be used in matrix or network stabilization by the application of a bias dependent upon the activity within the matrix.

Each element could, if required, include a self-biassing arrangement at input as illustrated in Fig. 4.9. This would be quite undesirable in microfabrication due to the size of capacitor needed for such auto-bias, but can be – and has been – incorporated in discrete circuitry built for trials purposes.

4.9 LSEs AS SENSORS

The circuit given for the basic feedback LSE has the appearance of a fairly typical oscillator circuit used in pulse and other waveform generators. Indeed, if a resistor is taken from input to supply, the circuit does become oscillatory.

Use can be made of this effect by, for instance, making the 'op–amp' component a phototransistor or using a photodiode in its bias chain.

The frequency of such a device is critically dependent upon the circuit constants as well as on the illumination conditions. However, as with so many of our requirements, this can become a positive advantage. In a latter chapter we shall find that the practical operational modes of pulsatory networks positively encourage the use of unmatched components such as photosensors which deliver unequal outputs.

4.10 ELECTROCHMICAL CELLS

The investigation of pulsatory cells would not have been complete without a series of experiments using electrochemical devices and components. A number of forms of electrochemical 'diode' assemblies have been tried with a view to their possible incorporation into electrochemical interconnection fields. However, this initial experimental work merely checked some ideas. If full advantage of the pulsatory systems is to be realised, a vast amount of work will be required in this field.

4.11 TEST PATTERN GENERATORS

Much of the original work in pulsatory matrices was conducted using an array of photo-sensitive components. Many ideas could be tested with relatively simple test rigs. However, for more detailed measurements, more standardized input fields were required.

Accordingly, sets of pulsatory delay elements were constructed such that an input line could be taken from each element of a delay line. Delay line lengths of 20 and 40 elements were chosen – but the details of the line could be determined by simple interconnections. This provided a rather flexible source pattern field which could be set into either random or stable fixed-pattern modes.

The generator elements used were simple-delays of the type illustrated in Fig. 4.10. Patchboards enabled the delay lines to be set in a variety of configurations so as to provide pulse streams to suit different experiments.

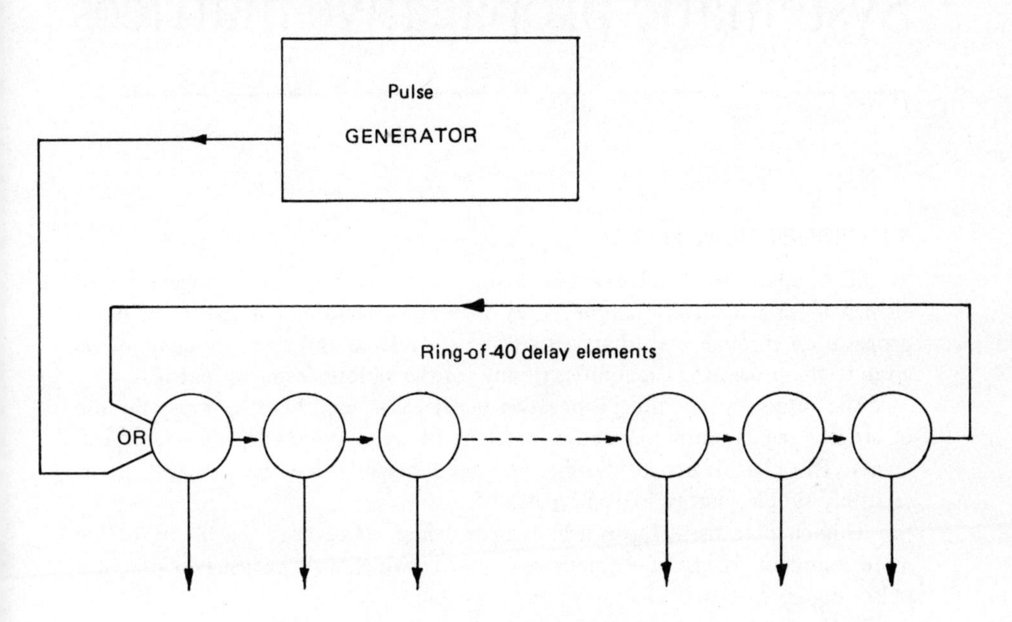

Fig. 4.10 – Test pattern generators.

5

Systematic propagative matrices

5.1 GEOMETRICAL MATRICES

In this chapter, we shall examine a number of geometrical arrangements of elements using relatively simple types of interconnection. The nature of signal propagation through such matrices will be considered and some thought will be given to the processing capabilities (if any) of the various forms of matrix.

The terms 'systematic propagative matrices' is used here to imply the use of strictly regular interconnection patterns between symmetrically spaced elements. The high degree of systematism used suggests that the elements are of relatively simple, that is, logical types.

This chapter then, forms a basis upon which to advance the theory of the more complex forms of element and matrix which will be used in practical pulsor matrix systems.

It is convenient to consider that each element has the same number of outputs as inputs. All outputs are taken from the same point in the elemental circuit. We can then express the element as having a fan-out equal to the number of inputs.

Initially we shall consider elements with three or four inputs. However, as we develop the concepts of interconnection specification and attempt to analyse signal flow, we shall sometimes require to base our analyses on 'inputs' and sometimes on 'outputs'. Elements both receive inputs and deliver outputs. These actions are not time coincident and thus can be considered as quite separate events.

The forms of interconnections which will be considered include:

(1) Direct forward connections.
(2) Divergent forward connections.
(3) Convergent forward connections.
(4) Simple plane bypass connections.

and of course, certain combinations of these.

Methods of launching information into a matrix will be considered and the two principal forms will be:

(1) Point-source with divergant propagation.
(2) Plane-source with divergent or convergent propagation.

As is quite commonplace in the design of electrical circuit networks, quite profound differences occur according to whether the outputs or the inputs of elements are taken as the underlying design considerations.

5.2 LOGICAL CUBIC LATTICES

Figure 5.1 illustrates the co-ordinate reference system used to denote the position of an element in a symmetrical lattice structure based on an isocline. In the course of this chapter we shall consider both point-source and plane-source signal injection.

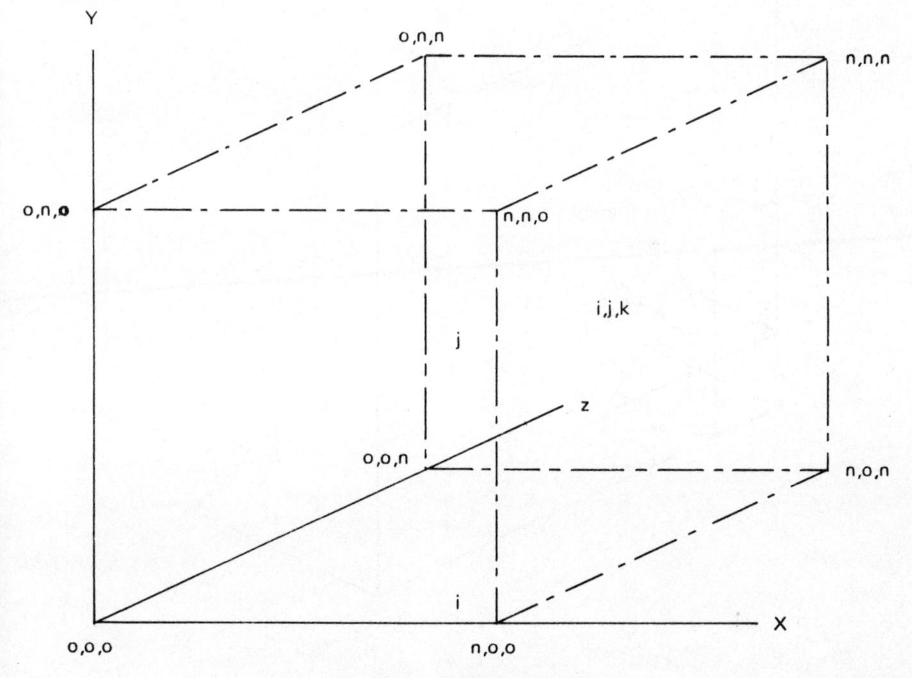

Fig. 5.1 – A Cubic Lattice.

The origin of the diagram (0, 0, 0) will be the source point for signal input in the one case. This arrangement, will give rise to a divergent form of propagation – if it does anything at all. In the case of a plane-source for the set of signals forming its input, the plane, (n, n, 0) will be the origin of the injected signals.

We shall assume the use of logical devices, most probably based on 'majority-logic' (2 out of 3, 3 out of 4 etc.) in the expectation of 'unity-gain' being unattainable at reasonable levels of pulse-probability.

5.3 THE POINT-SOURCE CUBIC

Given a single-line signal, it is reasonable to consider the possibility of using a point-source cubic lattice and question the nature of processing if propagation were to occur. Figure 5.2 shows a possible arrangement. Each element has three outputs coupled to three other elements at unit distances along the x, y and z axes.

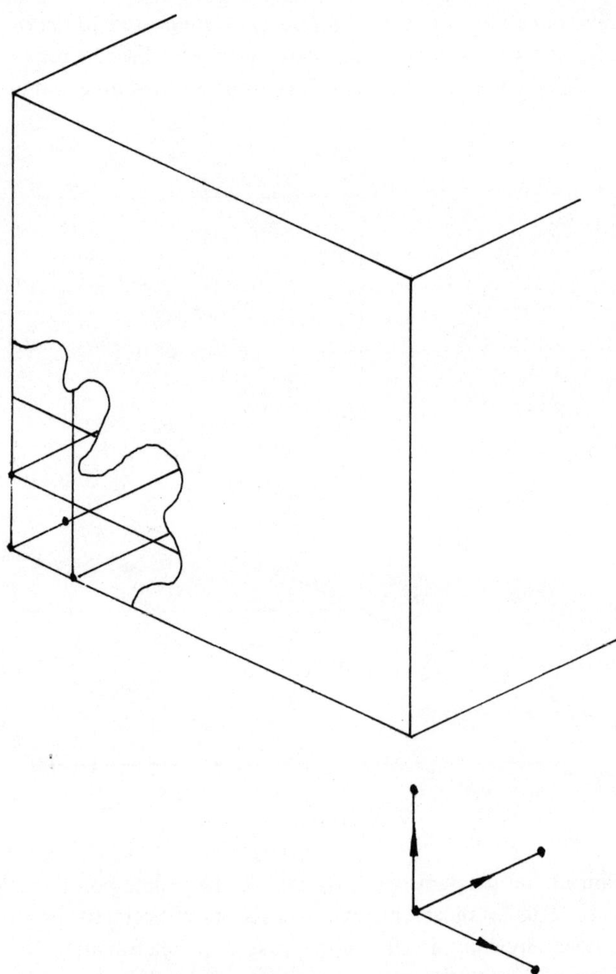

Fig. 5.2 – Triple-output Point-source Cubic.

Should the signal input conditions be such as to initiate propagation, it would appear that the propagation vector would be the principal diagonal of the matrix and that divergence would occur. However, the cubic geometry is non-simple in the central region of the principal diagonal. The later stages of the lattice would exhibit convergence of the signal(s). The output from the matrix would be at the remote apex on the diagonal.

It would be possible to extract versions of the input signal at points other than the output apex. The single-line input signal would not appear to contain information usefully processable at these intermediate facial or edge points on the lattice.

All-in-all, this matrix does not appear to offer the most exciting of approaches to pulsatory matrices.

Rather special characteristics are required of the pulsatory elements for use in the early stages of the matrix, that is, for elements at positions 0, 0, 0, 1, 0, 0, 0, 1, 0, and 0, 0, 1. Each of these receives one input only and hence may not be a logical element of input complexity exceeding that of OR. A sensitive LSE element could possibly have applications here.

All pathlengths from 0, 0, 0 to element i, j, k are equal. Thus, apart from individual elemental variations in response, all paths provide substantially equal delays. An 'event' in a signal would therefore propagate through the matrix to arrive at element i, j, k with some blurring in time.

Table 5.1 summarizes the conditions in the various elements of the point-source isoclinic matrix.

Table 5.1

Elemental Inputs – Isoclinic Matrix

Element Position	Number of Inputs (in)		
	1	2	3
0,0,0	Sensor		
1,1,1	X		
0,j,k		X	
i,0,k		X	
i,j,0		X	
i,j,k			X

Distance from 0,0,0 to i,j,k is i + j + k units of delay.

For unity-gain, sensitivity must be 'normal' multiplied by a constant which is inversely proportional to number of inputs.

5.4 PLANE-SOURCE CUBIC (QUAD-INPUT ELEMENTS)

Here, elements in the face-plane n,n,0 (Fig. 5.1) are coupled to adjacent elements of plane n,n,1 and so on throughout the matrix. The boundary conditions here differ from those of other lattices examined to date. Propagation from a central element of the source-plane would appear to be simply-divergent until the lattice boundary is reached. The principal direction of propagation is normal to the source-plane.

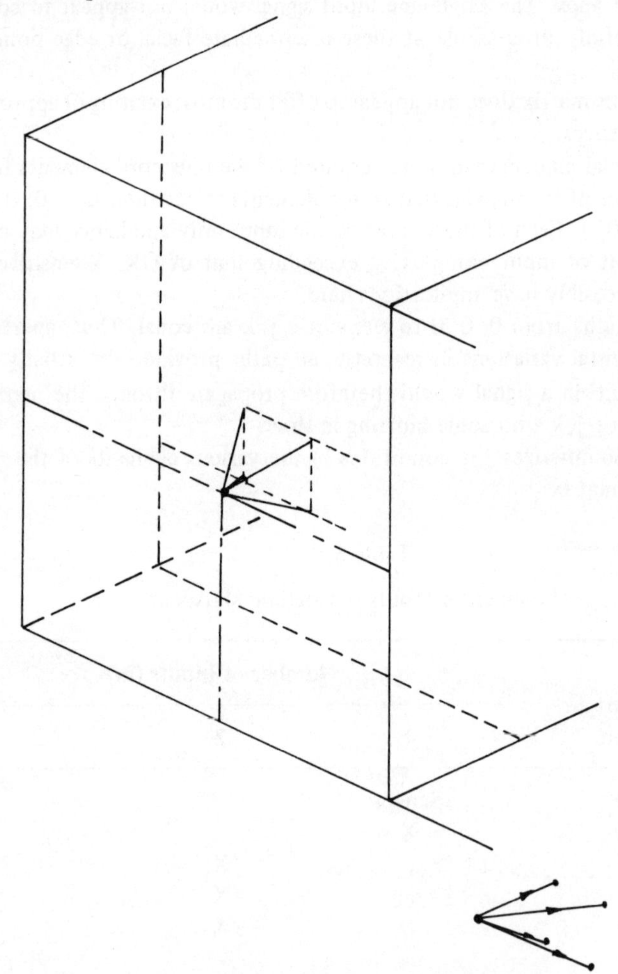

Fig. 5.3 – Quad-output Plane-source Cubic.

The 'processing' characteristics of this matrix are not, however, attractive. It would seem that if individual 'signals' were applied to particular source-plane elements, there would be an integration effect during any propagation which did occur. In later chapters we shall examine techniques which enable us to design far more useful matrices with true processing capability.

5.5 DIVERGENCE, CONVERGENCE AND BIVERGENCE

Point-source analysis of the point-source orthoclinic has shown it to be propagatively divergent over part of its path network and convergent over other parts, culminating in a single principal output. The plane-source tetrahedron however is convergent.

When consideration is given to the nature of any 'processing' that our simple matrices could perform, we find that there is in general a 'blurring' of the image. In certain systems this may be a desirable feature – the tetrahedral optical control system was an example. The general finding however is not encouraging.

Given a very simple 'signal' which could move across the face of a plane-source cubic, any image formed on the final plane would be heavily blurred but may well have a distribution such that the position of the signal on the face-plane could be deduced from the distribution of signals at the output plane. Were we to design a matrix to produce a roughly Gaussian distribution of a simple signal when that signal were centred on the face-plane, the distribution at the output plane could well tend to Poisson as the signal traversed from the centre of the face-plane.

The 'bivergence' (combined divergence and convergence effects) found with certain matrices suggest that there may be ways to overcome some of the more obvious disadvantages of the simpler interconnection fields. It is not considered appropriate to encumber this text with the full details of analyses which have been performed on numerous types of matrix. We shall however, consider just a few of the underlying principles.

5.6 TRIPLE-INPUT PLANE SOURCE TETRAHEDRAL MATRIX

A version of this device was analysed in Chapter 2. There, computations indicated that a surprisingly large number of system parameters could be derived from the single-line output from this matrix. In fact, much of the early theoretical work on pulsatory matrices was performed on the tetrahedral format. Also, an optosensory plane input tetrahedron was designed, constructed and tested using the rather primitive transistors of the early 1960s. It was in part due to the indications of potential processing power of pulsatory matrices resulting from the tests that the decision to proceed was taken. The rather unlikely-looking tetrahedral format does then at least deserve its place in the history of our development work even though it hardly ranks as a practical device.

Figure 5.4 gives a stylized sketch of the matrix form and the inset illustrations the manner of geometrical layout of the element interconnections used

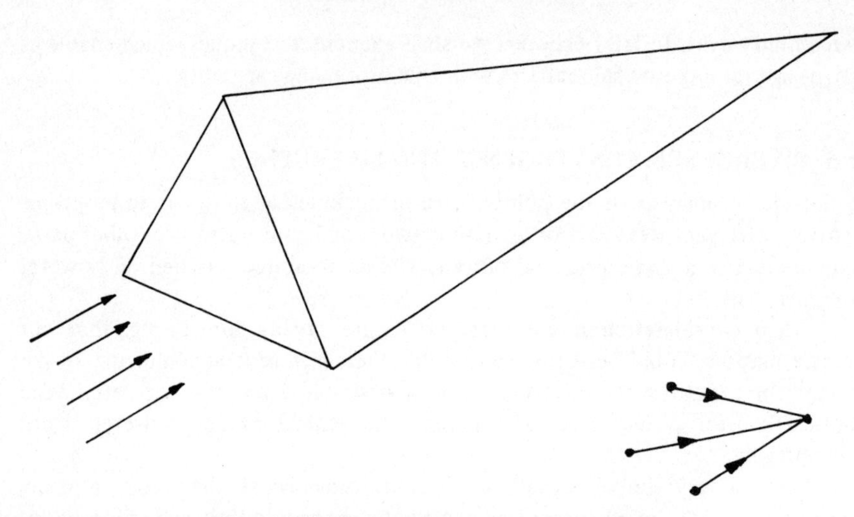

Fig. 5.4 – Triple-input Plane-source Tetrahedron.

throughout the matrix. The title of 'triple-input' was used because of the naturally convergent nature of the matrix.

5.7 TRIPLE-OUTPUT POINT SOURCE TETRAHEDRON

This device ranks with the point-source cubic lattice in that it would appear to have little to offer in practical processing even if only because of the simplistic nature of the input signals it could handle. It is included here as a format which should at least be mentioned in a general discussion of lattice arrangements.

A stylized sketch of the matrix and inset of the interconnections plan appear as Fig. 5.5.

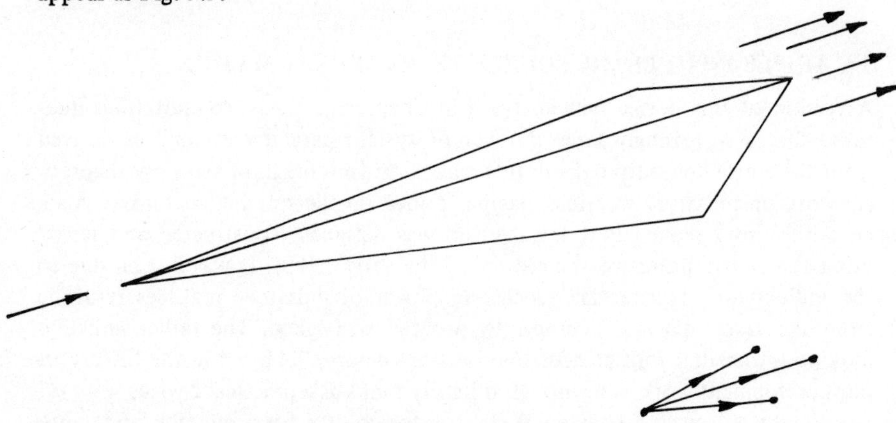

Fig. 5.5 – Triple-output Point-source Tetrahedron.

5.8 QUAD-INPUT TETRAHEDRON WITH PLANE BYPASS

Using the structure described earlier for triple-input elements, we could add a fourth input and feed it from two planes removed using an element in a corresponding axial position. Thus an output pulse will be applied to three elements in the next plane (these being spacially displaced in an interlaced manner) and also to the element two planes bypassed in a similar axial position. There is then a form of 'anticipation' of the signal by delay minimization.

An 'event' in the input signal will then reach the output in the form of an increasing amplitude until maximum response obtains. Signal decay would follow a similar pattern.

The degree of modification of the signal contour in time will be dependent, among other things, on the 'length' (number of planes) of the matrix.

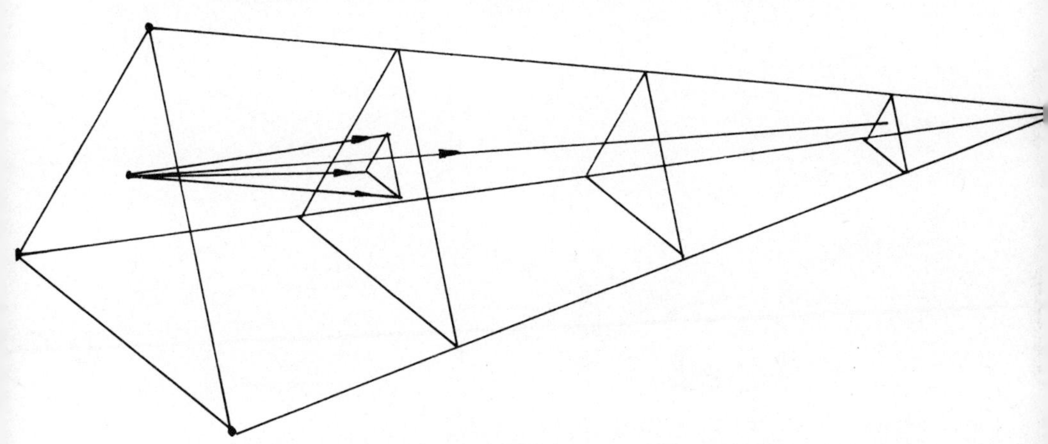

Fig. 5.6 – Quad-input Tetrahedron with Plane Bypass.

5.9 PLANE-SOURCE TUBULAR MATRICES

There are four principal forms for a tubular propagative matrix: based on their cross-section geometries:

(1) Triangular.
(2) Rectangular.
(3) Hexagonal.
(4) Polyhedral approximating to circular for fairly large matrices.

In all cases, the boundary conditions of the matrices constrain the form of propagation. Simply-divergent and simply-convergent descriptions cease to be meaningful. Questions such as 'Should the boundary elements be of different type from the in-bound elements?' immediately spring to mind. It is at this stage that certain forms of complexity arise which place limits upon the amount of study that should properly be applied to these forms. Conditions pertaining to the reliability of such devices were they to be constructed cause one to think seriously of abandoning this line of enquiry. Suffice at this point to note that these geometric forms will reappear in highly active and extremely reliable forms in later work with probabilistic interconnection forms using LSEs.

The triangular-section arrangement of Fig. 5.7 is very basic. It will probably prove too restrictive for production systems. However, other more complicated figures may be fabricated from triangular sections. The principal objection to this form is its lack of symmetry in that there is a considerable variation in centre-to-edge measure over a range of angles. The triangular format does not accord closely with naturally-occurring signal sources as we presently know them. The result is that under transformation such as signal rotation, there would be considerable mismatch between signal and sensor/matrix.

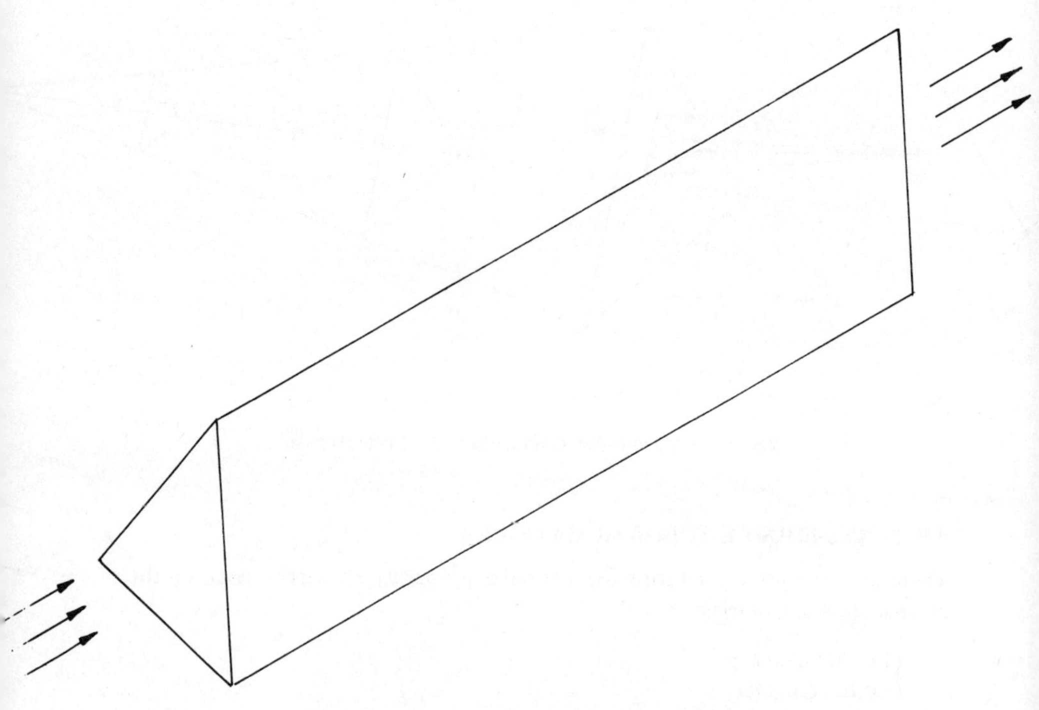

Fig. 5.7 – Triangular-section Propagative Matrix.

The rectangular section of Fig. 5.8 would appear to have some advantages over the triangular but the gains do not appear to be great. The concept of 'bundle' constructions of large matrices leaves the rectangular sectional form with a number of problems – should edges be aligned or overlapped? How should one deal effectively with a rotating signal?

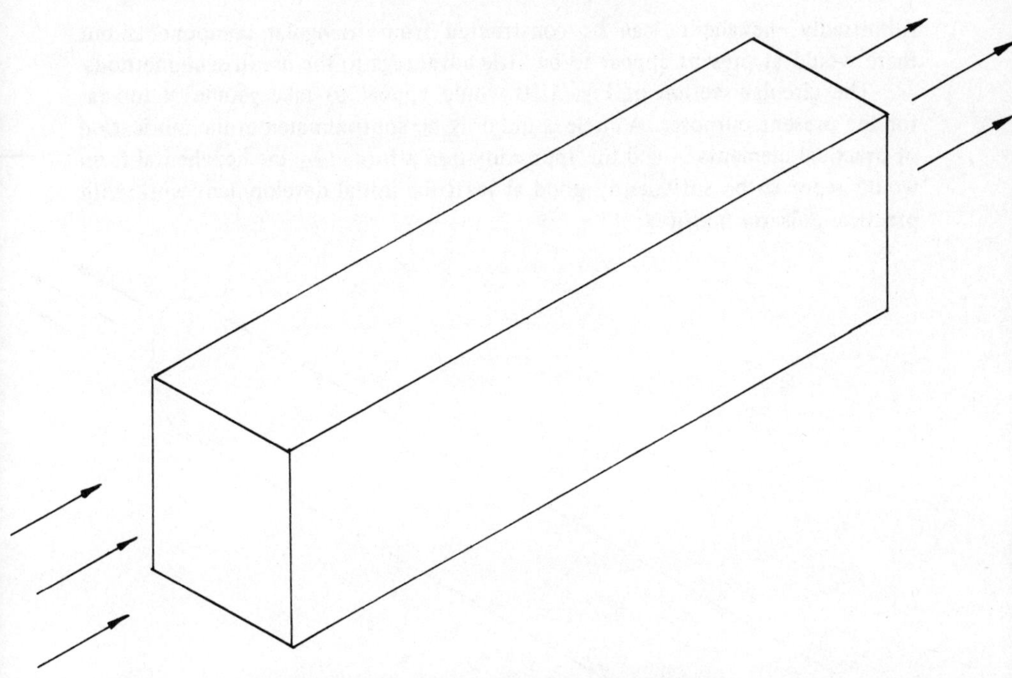

Fig. 5.8 – Rectangular-section Propagative Matrix.

For a 'bundled' form of fabrication, the arrangement of Fig. 5.9 would appear to offer considerable advantages over the other forms introduced in this chapter. Sets of hexagonal sectioned tubes will pack closely – Automation of silicon-based device production favours regular arrangements such as these.

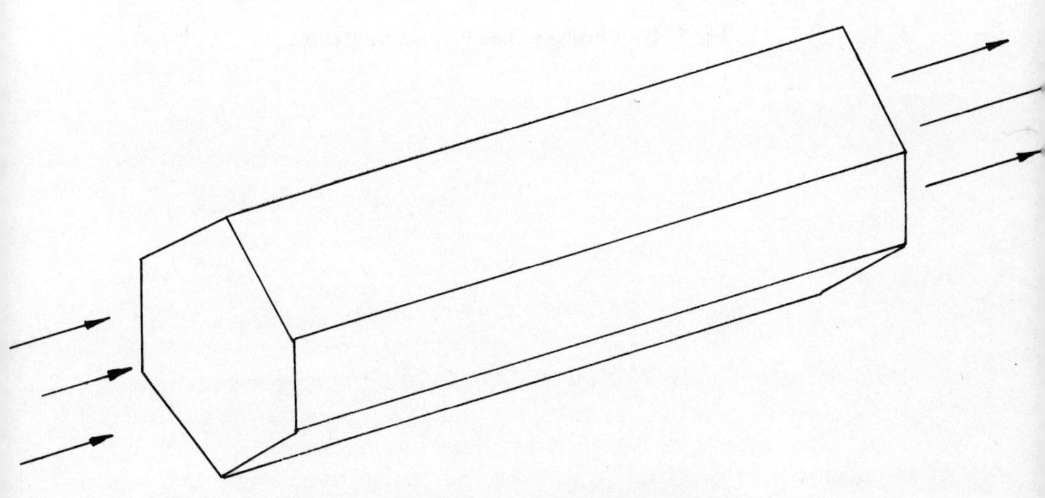

Fig. 5.9 – Hexagonal-section Propagative Matrix.

Admittedly, hexahedra can be constructed from triangular components but there would at present appear to be little advantage to the use of such methods.

The circular section of Fig. 5.10 would appear to take geometry too far for the present purposes. A circle could only be approximated in the fabrication of practical elements – and the approximation afforded by the hexahedral form would seem to be sufficiently good at least for initial development work with practical pulsator matrices.

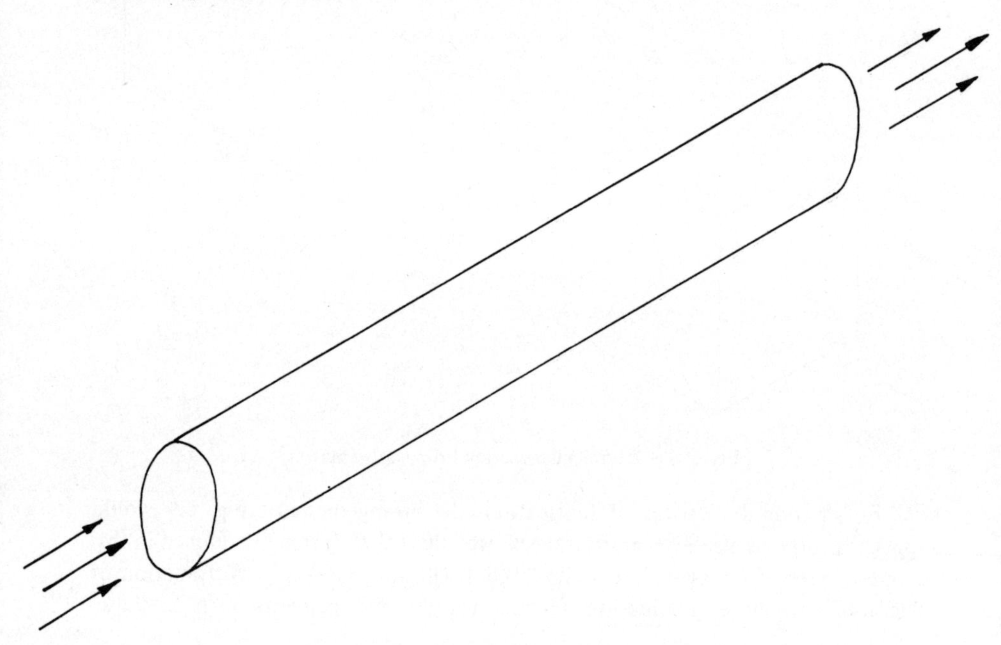

Fig. 5.10 – Circular-section Propagative Matrix.

6

Application of the Positive Principle

6.1 TOWARDS RELIABILITY – DESYSTEMIZATION

We have seen that the simply-connected, systematic matrices possess limited processing power and that they have great sensitivity to component failure. One could reason that the use of redundancy should improve the situation. The conclusion reached in Chapter 1 suggests that this is not however a satisfactory approach. We search for a scheme in which interconnection failures as well as component failures will not adversely affect overall system performance.

Improved processing power may be obtained by other than direct plane-to-plane interconnections. There are many possibilities for interconnection schemes which are other than those indicated in earlier chapters. A given element could be coupled to planes beyond the adjacent plane, it could be coupled to other elements of its own plane or it could be coupled to elements of preceding planes.

Forward-coupling to more remote elements would provide a form of reduced propagation time which would be signal-dependent. This can be likened to some forms of servo control stabilization technique which are well understood. Backward coupling could be used to provide forms of selective signal enhancement and lead to the development of certain types of memory system. In-plane coupling can be used to induce sensitivity to and control of certain image characteristics such as the sharpening of fuzzy edges.

The LSE was selected as a fundamental processing element by virtue of its inability to store data. It was suggested that if this 'memory element' could not actually hold data then a system which incorporated large numbers of such elements would not necessarily fail were a single element to fail. The claim that this could be so went unsubstantiated in earlier chapters and was not upheld in any of the matrices to which it was applied in earlier chapters.

The principle of Mathematical Induction suggests that what we have done once we can do again. We shall thus apply the same concept to the fields of interconnections between elements. The goal is to make system performance independent of interconnection failures.

All the systematic interconnection fields examined to date share the characteristic that the behaviour of a matrix is uniquely determined by the form of interconnection fields used. Any failure in a systematic interconnection field results in degraded performance.

6.2 DEFINITION OF TERMS

The two principal ways of describing interconnections are:

(a) Supply signal TO a device.
(b) Obtain input FROM a device.

In general we shall consider that the purpose of an element of matrix is to provide signals TO other devices.

6.2.1 Plane Bypass (or Penetration)

The technique of forward-projection to elements in remote planes will be termed 'plane bypass'. Another way of looking at the concept is that the output from a particular element will 'penetrate' forward into the matrix. This is a fundamental principle in the design of pulsatory matrices (or Networks) and will receive considerable attention throughout the remainder of this book.

6.2.2 Backcoupling

When a small proportion of the inputs to an element are obtained from elements in a succeeding plane, the term 'backcoupling' will be employed. It will be found that backcoupling provides a small degree of image enhancement and will become a normal feature of the advanced forms of matrix.

6.2.3 Reversed Elements

When a number of the elements of a matrix are coupled to preceding rather than succeeding planes, new forms of image enhancement appear. The degree of enhancement achievable extends from the single forms provided by backcoupling to the appearance of actual image memory of well-defined types.

6.2.4 Intraplane coupling

This will most commonly be coupling between local elements of a plane and the most usual reason for its use will be static enhancement rather than the space-time dynamic enhancement and other image processes provided by backcoupling and reversal of elements.

All four forms of interconnection will be found in the matrices which follow.

6.3 DESIGN CONSIDERATIONS

Just as we approached the design of an element by using probability theory, so we can approach the design of interconnection fields. The premise is that the interconnection field must provide sufficient interconnections to render the matrix active and, combined with the characteristic behaviour of the elements, must cause but trivial effects on overall performance when there is a failure in either an element or an interconnection.

If systematic interconnections lead to poor reliability, we should not use such 'wiring specifications'. Instead, our 'wiring list' should indicate that "some of these elements should be coupled to some of these elements.". This is against the normal principles of electrical and electronic design. There are however, precedents. The Conditional Probability Computer (NPL 1960), the Homeopath of 1950 and other devices, have shown 'self-repair' by the establishment of alternative signal routes is feasible.

6.4 INTERCONNECTION SPECIFICATION

The 1966 patent of the author introduced the concept of the probabilistic interconnection field. In essence one can explain its operation in such terms as 'if we do not know the wiring because there is no regular wiring scheme, then provided we keep within the statistical definitions of the wiring requirements, whether or not we connect specific elements is immaterial to the final result'.

Our wiring scheme then is that a statistically-expressed quantity of interconnections shall be made within a given region of the matrix. Our specification for interconnections must quote the range and distribution of interconnections throughout the matrix space.

A list of interconnection requirements is given in Table 6.1. By careful design, one can determine the behavioural characteristics of a completed matrix. In fact, the term 'matrix' is no longer applicable to the curious interconnection field which results. The term 'network' will therefore replace 'matrix' in respect of the probabilistically determined interconnections fields.

To satisfy the requirements of fail-safe, quite large numbers of elements are needed. Microfabrication eases this problem. The resulting networks will certainly demonstrate that performance is no longer determined by elemental or interconnection failures.

Table 6.1
Interconnection Specification

Parameter	Example
Number of elements	some thousands or millions
Inputs per element	Mean (say 20), Distribution (say equiprobable or Normal, $\sigma = 5$)
Plane bypass	mean (say 10), Distribution (say Poisson, $m = 4$)
Backconnections	say 2%
Intraplane connections	say 2%
Reversed elements	from say 1% to 10%
Lateral spread (divergence)	say ± 3, equiprobable distribution

Many tests have been run on such network specifications using computer simulation (Chapter 12). The concepts introduced in this chapter find very strong support in all these tests.

6.5 PLANE BYPASS CONNECTIONS

Figure 6.1 illustrates the inter-plane connections used in some of the preceding work. The concept of plane bypass is shown in Fig. 6.2. Here the mass of interconnections have been stripped away to leave the essentials of signals from one element.

The stylized diagram of Fig. 6.3 illustrates the concept of the distribution of numbers of interconnections against plane bypass values. A natural form of distribution for such a system is the classical Poisson distribution. Figure 6.4 takes the information of Fig. 6.3 to a more useful degree of abstraction leading directly to the adoption of the Poisson approach. The formula for the Poisson distribution is:

$$Qn = \mathrm{e}^{-m} \left(1 + \frac{m}{1} + \frac{m^2}{2!} + \frac{m^3}{3!} + \ldots + \frac{m^n}{n!}\right) \times T$$

Tables of this function are to be found in most statistical textbooks. We have simply taken the ordinary Poisson Series and multiplied it by the total T number of inputs to a given element, T being the design maximum. Each term of the expression then gives the average number of connections to be made to elements at a given (the nth) plane.

The value of a term from a Poisson series is non-integral. However, this does not mean that we cannot make the exact quantity of interconnections required by the series. The value is a statistical quantity and its interpretation must be that the average quantity of such interconnections is to be the term value. No detail is given or implied about the connections to or from a *given* element.

Given that a network of interconnections is simply-Poisson connected, that is no backcoupling, no intraplane connections and no reversed elements, we can compute the effects on the network of a change in the established quiescent (or ambient) excitation field.

Consider that a given simply-Poisson connected network has a sufficient ambient excitation to raise its sensitivity level to the point at which a simple change such as an 'impulse' would propagate through the network, it is readily seen that the impulse will be sensed at the output as a result of the deepest penetration of the plane bypass connections. The output level at this time will be low. However, as the signal propagates further through the matrix network, the output amplitude will rise following fairly closely the reflection of the poisson distribution of the interconnections.

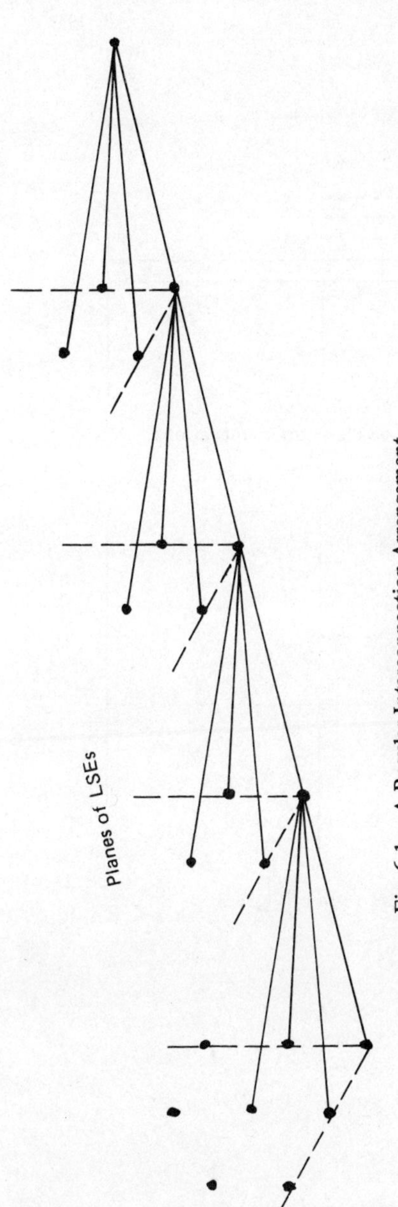

Fig. 6.1 – A Regular Interconnection Arrangement.

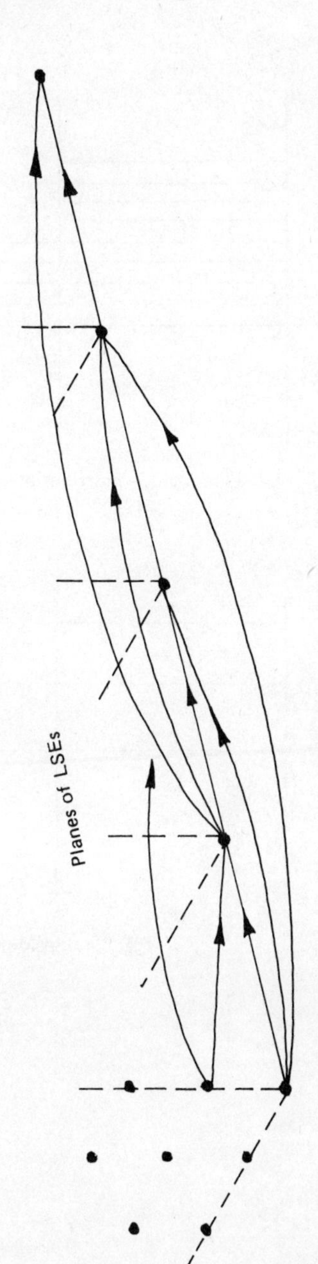

Fig. 6.2 – Plane Bypass Interconnections.

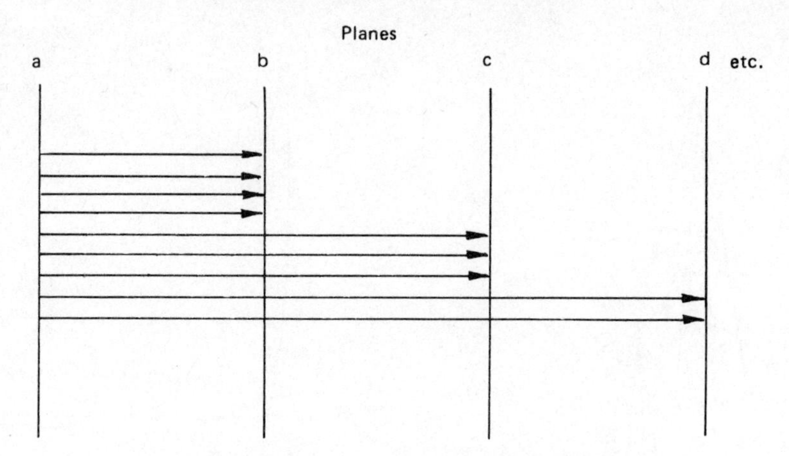

Fig. 6.3 – Distribution of Element-to-Plane Interconnections.

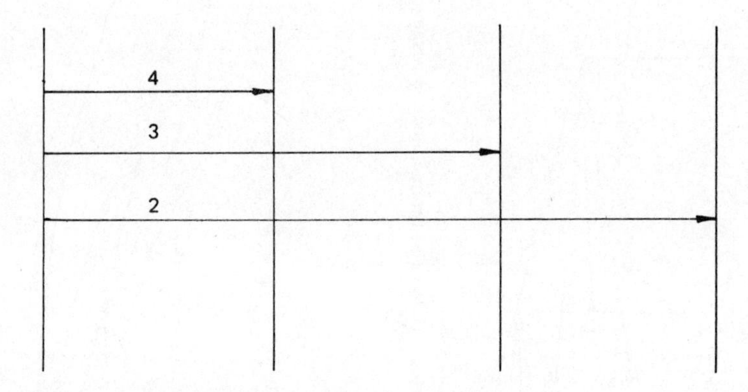

Fig. 6.4 – A More Convenient form of Fig. 6.3.

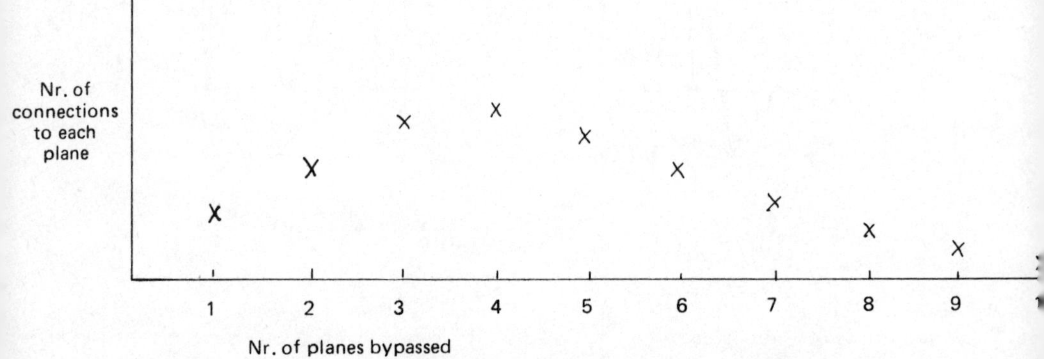

Fig. 6.5 – Use of Poisson Distribution to Plan Bypassing.

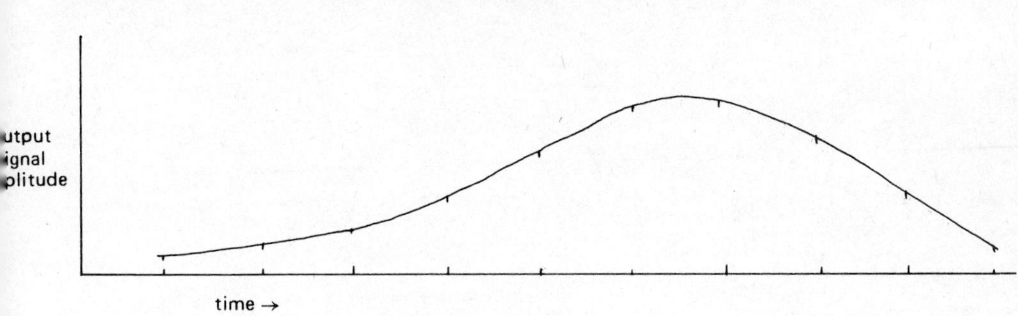

Fig. 6.6 – Network Output in Response to an Impulse.

6.6 PENETRATION AND DIVERGENCE (OR SPREAD)

Just as the depth penetration of the output from an element is specified statistically, so must the lateral spread. This parameter of the network may be assigned Normal or Equiprobable distribution about some mean or within a given range of element 'bypassing' (see Fig. 6.7).

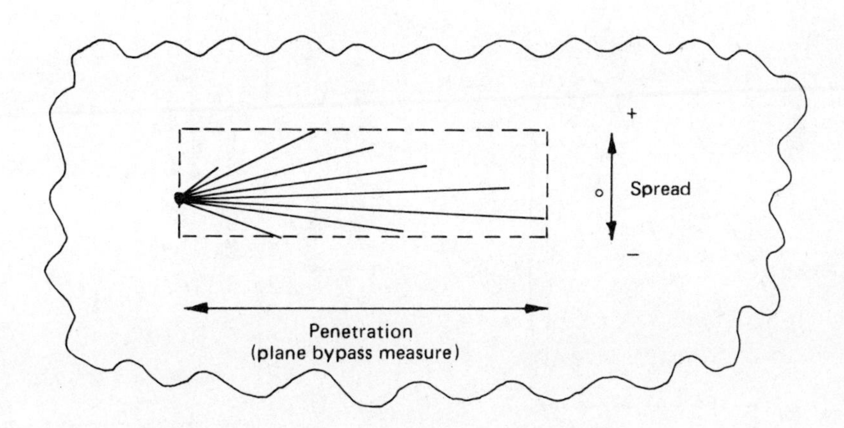

Fig. 6.7 – Defining Penetration and Spread of Interconnections.

6.7 BACKCOUPLING

Figure 6.8 shows tha concept of backcoupling. The normal form of connection is from an output to an element in a succeeding plane. In practice, a small percentage of connections would be backcoupled.

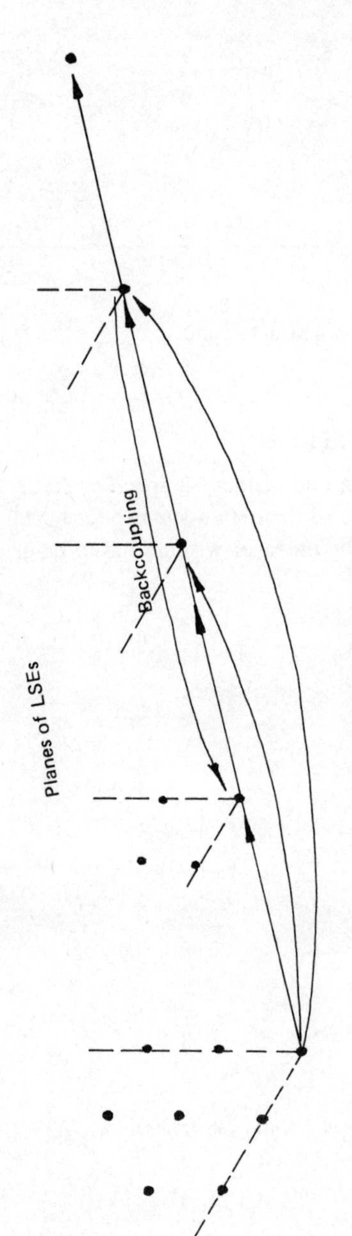

Fig. 6.8 – The Backcoupled LSE.

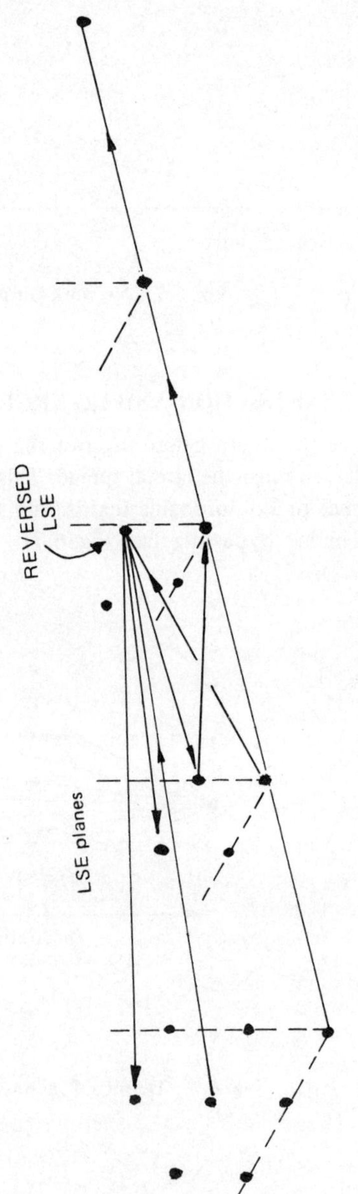

Fig. 6.10 – The Reverse-connected LSE.

6.8 INTRAPLANE INTERCONNECTIONS

The intraplane connection arrangement is illustrated in Fig. 6.9. As with back-coupling, a small percentage of such connections would normally be made. The effect will be either enhancement of 'grey' edges or more positive discrimination in favour of edges according to the nature of the input to which connections are made.

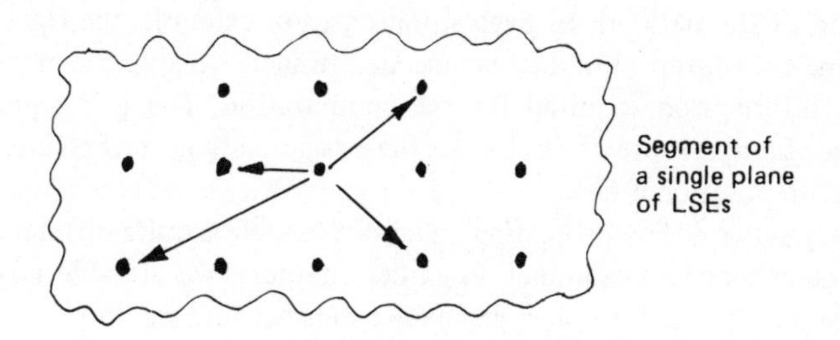

Fig. 6.9 – Intraplane Interconnections.

6.9 REVERSE- CONNECTED ELEMENTS

In Fig. 6.10 the nature of interconnections of reversed elements is outlined. It will be readily seen that this can lead to a curious form of image enhancement approximating to localized memory. It represents a form of 'distributed memory' region capable of being excited into a stable representation of a given form of signal.

Such 'memories' can be established at different distances along the network according to both network conditions and the nature of the excitation.

The design of such networks then, is a matter of deriving base parameters from the sets of (now exceptionally complex) fields of probabilistic equations. In this work, the computer modelling techniques of Chapter 12 have played a significant role. Confirmation of the underlying premises of pulsormatrices and of pulsornetworks has been obtained. In particular, during the numerous simulations which have been run over the period from 1975, the validity of element independence has been established.

6.10 THE NETWORK TRANSFER FUNCTION

This is, of course, quite a complex topic involving time-space behavioural characteristics. Analysis would appear to be hampered by lack of information on the individual states of elements, the details of the interconnections fields and by the variety of the types of 'signal' to be dealt with.

Commencing with the 'signals', they will all be at least two-dimensional fields of information varying in time. The characteristics of both elements and interconnections are known only statistically – and we have insisted that we should not enquire into the actual details of any element or any actual interconnection. The performance must not be significantly influenced by failures in any component of the network. How then can analysis commence?

Assuming that a network has been constructed, what do we know of it? Actually, rather a lot. And the information is rather encouraging. The statistical characteristics of the network details can be used directly to compute the reaction of the network to given situations. For example, the Gain profile of the network to a given stimulus can be determined because we have all the parametric information required for the computation. The gain equations become less specific and more general – so there is actually no real change in the nature of the computational task.

Our network has processing capability which is quite different from that of the regular matrices examined in earlier chapters. We are able now to compute the response profiles to changing excitations because we know the characteristic properties of the (now very complex) interconnections field.

All in all, the task of network design and assessment now appears to be a rather more straightforward process than would at first sight have seemed. In the absence of backcounting, intraplane interconnections and reversed elements, the task of design reduces to the choice of a number of parameters such as:

(a) the trigger level of the elements
(b) the decay constant
(c) the complexity of interconnections (number of inputs per element)
(d) the penetration of plane bypassing used
(c) the divergence (lateral spread).

Given a sufficiently large matrix to deal with the source information field, then the elemental and network parameters indicate the type of behaviour to be expected.

In general, the gain of a network will increase according to:

(i) complexity
(ii) elemental sensitivity

and will decrease according to:

(a) spread or divergence
(b) decay.

There will be pattern sensitivity according to the gain and other parameters such as penetration and divergence.

When other factors such as interplane interconnections, backcoupling and reversal are introduced the effects range from image enhancement to actual memory of images.

All of the parameters referred to will affect the resolution of the propagated image and the velocity of propagation. If we can discriminate between the decay of an image as it propagates (due to inadequate gain) and the increased propagative action of a gain-enhanced network, then we shall find in general that the velocity of propagation is proportional to the penetration and is adversely affected by the feedback techniques.

When a reasonable proportion (say 6–8%) of the elements are reverse-connected, then stationary images can be set up in the network. This provides discriminatory action in favour of impressed images against non-sympathetic images.

In certain cases, we shall find that the outer surfaces of networks contain information relating to the movement of information across the image source-plane (the input plane). Thus we can see many possible applications to the realm of image processing. Our networks can be made tolerant of many changes in the external environment and of failures of components within the network.

The reasoning is quite simple – if we do not know and cannot determine the 'purpose' of an element because it is just one item in a statistically-described population, then it will not matter if that item is absent.

Whilst it has not yet been possible to construct networks with many thousands of components, it has been possible to simulate them on computers. Many simulations have been run using a wide range of parametric descriptions to which the computer had to work in 'constructing' the networks. Most such simulations have also specified the actual existence of defective elements. In many cases, whole rows of elements have been defined as defective and the simulations have confirmed that the effects are trivial.

6.11 INHIBITORY FUNCTIONS

The gain locus of a network is highly non-linear. Unity-gain is at a critical point on the locus – and that point is highly signal-dependent. Some forms of gain stabilization are obviously required in a practical device.

In electronic systems, control may be exercised by application of biassing potentials and the like. Control is also exercised by use of feedback and feedback control. We could extract from the network a signal or set of signals representing the activity levels present and use these as stabilizing factors. Another possibility is the use of actual inhibitory signals.

Were we to arrange that the count of an element be inhibited by some means – over-excitation or a discounting input, then control of network parameters may be achievable.

In fact, by arranging that a small proportion of the inputs to elements will cause the count to be blocked then, as activity rises, so the number of elements whose count had been affected would rise. The net effect is that at the higher levels of activity, the gain would actually be reduced – a most satisfactory state of affairs. This form of behaviour has been included in the computer models of pulsornetworks.

A number of media exist for the setup of suitable interconnection fields. Certain sets of connections could be fabricated on the substrate of the micro-electronic devices. Other types of interconnection with a variety of characteristics can be formed by electrochemical means – greases, static or pumped fluids and so on. Yet another device which can offer very special facilities is the fibroptic bundle. Together with the transmission characteristics covering single-channel communication, a fibroptic bundle can provide sets of plane-to-plane mappings and plane-to-line mappings. Such a bundle can therefore be used not only to interconnect, but to alter the geometric organization of the source field to suit some foreign geometry in the receiver device inputs. The fibroptic bundle does place limitations on the resolution of images which it can transport but is nevertheless, a device which is a fair match to the sizes of microelectronic element required for pulsor matrices.

We are then, able to consider the design of interconnection fields to suit a considerable variety of coupling and translation functions between very large numbers of active elements. Practical pulsor devices are feasible with very high packing density, greatly exceeding that of present-day devices and with very large quantities of summator elements (exceeding say 10^8). We may expect the functional reliability of such devices to be very high because we have decoupled the relationship between individual component failures and system failures.

7

Dynamic processing matrices

7.1 PROCESSING BY LSE PULSATORY MATRICES

Here we shall consider some of the processing capabilities and characteristics of these forms of matrix. The work of Chapter 6 is the start-point. The nature of the responses of pulsor matrices to stimulating signals will be considered in some detail and use will be made during this investigation of a number of the conceptual stabilization techniques considered briefly elsewhere.

The three-dimensional nature of the matrices makes possible the handling of complex two-dimensional images varying in time. For more simple planar matrices, single-dimensional signal sets could readily be handled. However, as the complexity of our matrices develops in the remainder of this book, it will become clear that the potential processing capability of fully-developed pulsor matrices exceeds that of computer systems by many orders of magnitude.

For the present, we shall take the class of matrix developed in Chapter 6 and attempt some analysis of its processing potential.

7.2 THE INFORMATION CLOUD

The response of a pulsor matrix to a stimulus is the setup of a 'cloud' of triggering responses. The nature and behaviour of the cloud is dependent upon the stimulus, the nature of the matrix and the state of the matrix. The state is continually changing – the matrix is a dynamic device.

In this chapter, we shall examine the setup of 'information clouds' and trace their progress in relatively simple LSE matrices. This study will disclose some of the dynamic processing properties of simple LSE matrices and prepare the way for a study of more complex types of matrix.

When a group of adjacent elements is excited into trigger state by an input 'signal', other elements in the locality receive inputs resulting from the triggering of the excited elements. Because the triggering elements are close in space, certain other local elements will receive inputs from more than one triggering element.

It may be that during an initial excitation of a group of 'sensor' elements, certain other local elements will be raised to trigger level. At least, a number of such elements will be raised to near trigger state. Then a subsequent trigger would complete the task of triggering other local elements. Fairly rapidly, the various elements local to the stimulus will respond and incite propagation of the signal. Figure 7.1 illustrates the effect and indicates the 'seeded trail' in the wake of the travelling cloud.

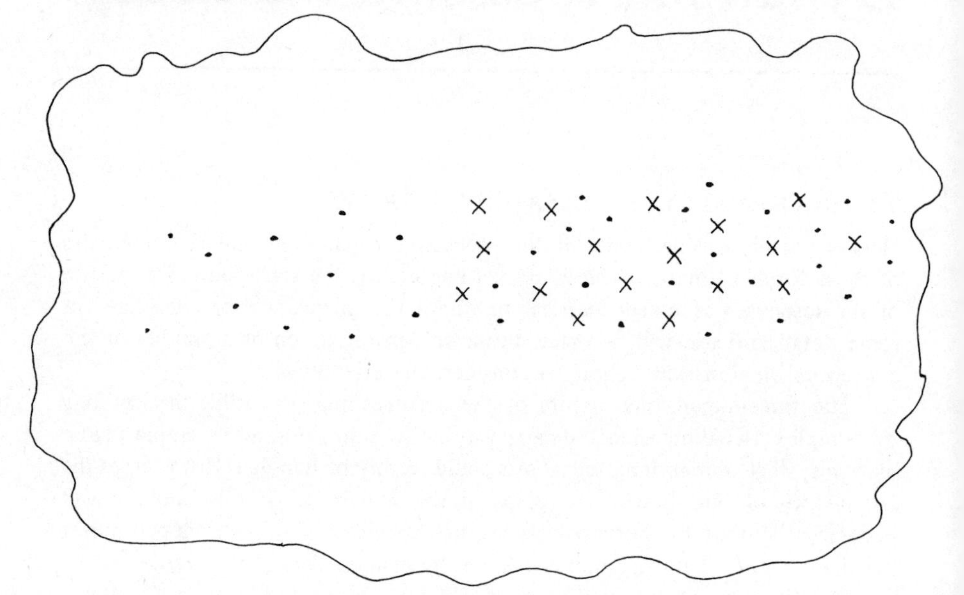

Fig. 7.1 – Matrix Activity and the Seeded Trail.

On cessation of the input signal, there would be no further sensor activity and the signal in its new form would 'detach' from the sensor 'plane' to propagate along the matrix according to the somewhat complex set of rules of these matrices. The three principal propagation characteristics are:

(a) The information 'cloud' would die, having travelled but a short distance.
(b) The 'cloud' grows to 'flood' the matrix.
(c) The 'cloud' forms as a stable entity and propagates satisfactorily.

7.3 PROPAGATION VELOCITY

This is dependent upon many factors, amongst which are:

(a) intensity of stimulation
(b) 'gain' figure (or locus) for the matrix
(c) state of excitation of the matrix
(d) effects of any stabilization techniques used.

The gain locus is itself dependent of the matrix configuration. Thus the nature of the distribution of interconnections in x, y and z directions is a fundamental factor in the determination of cloud velocity.

For example, velocity would apparently increase if the depth of penetration of interconnections (plane bypass) were increased. Taken alone, this increase in the volume covered by the output connections from a given element would merely serve to reduce the gain and hence reduce the probability that a cloud would be supported.

Without some form of gain stabilization then, the conditions under which propagation could be supported would be extremely critical. Many stimulations of such situations have been run and the results bear out the conclusion that clouds will in general either decay or 'explode'.

Much of the following reasoning depends upon the ability we have to design stable matrices.

7.4 DETECTION OF MOVING STIMULI

Given that an information cloud can form and propagate in a stable manner through a given matrix, consider the case of a stimulus which moves across the 'face' of the sensor 'plane'. The cloud within the matrix will effectively trace the path of the stimulus moving along a vector given by the propagation velocity along the matrix axis and the transit velocity across the major axis.

It would then, be possible to detect the movement of the image by a 'casing' of responsor elements around the matrix. The required information is present in the matrix and can readily be extracted and reduced, for example by further specialized LSE matrices. In fact, the impression is rapidly gained that a full-scale processing matrix would take the form of 'bundles' of elements having propagative and communicative features.

This suggests that a particularly useful form of bundle would be the hexagonal-sectioned tube which tends to optimize the use of space in a structure and also provides a number of surfaces for inter-bundle communication.

7.5 TIME-SPACE DIFFERENTIATION

A signal which varies in time may be detected and analysed by the pulsor matrix in any of a number of possible ways, a change in signal level will be magnified during transit through the matrix. Use of the auto-stablizing effects of inhibition and parameter control will minimize the selectivity and hence the detection of signal changes. This is akin to automatic gain control (AGC). However, it is quite feasible that with some forms of elemental circuit, the onset of either inhibition or parametric control can itself be used as a differentiated output response.

Space differentials may be obtained, for example by use of the inter-bundle communication set up by propagating signals. Pulsatory matrices have a remarkable facility for signal and event discrimination. As a result, designs can be foreseen in which both time and space differential information can be extracted and used in control and processing of complex signals.

7.6 THE ROLE OF INHIBITION IN AUTO-STABILIZATION

Use of a proportion of inhibitory effects would have little consequence when matrix activity was low. Increasing activity would raise the probability of activating inhibitory actions. Thus, some degree of gain-limiting would be provided. Figure 7.2 illustrates the effect of high and low pulse input rates. A very high drive rate causes inhibition of output.

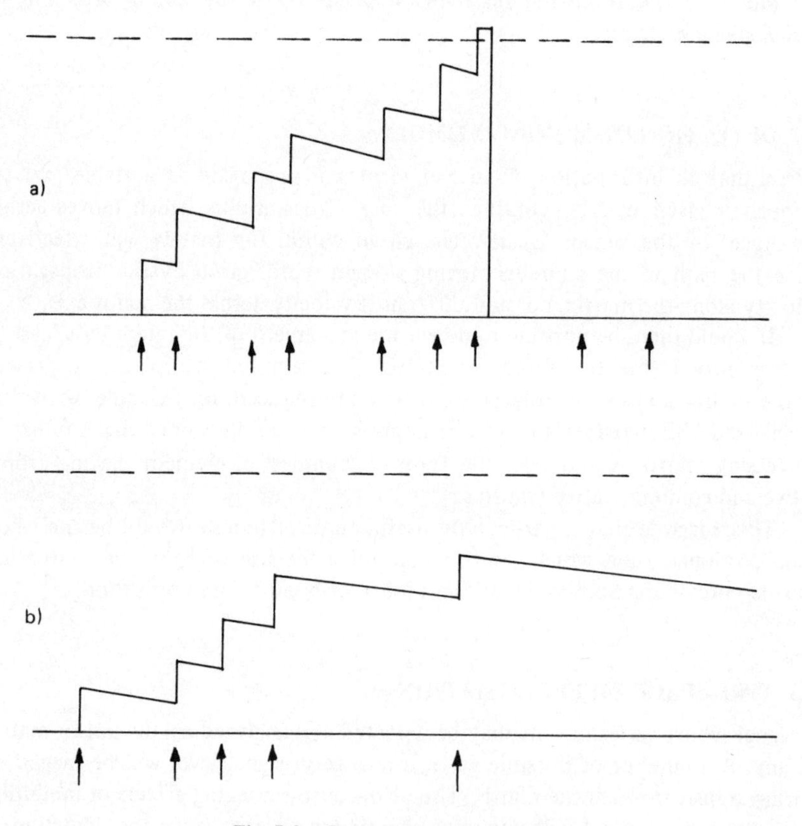

Fig. 7.2 – Pulse Rate and Inhibition.

The effect is brought into action simply by increased local activity. It is a purely localized activity and hence would appear to provide just that form of auto-stabilization required. Unfortunately however, this is not a sufficiently pronounced effect to result in full stabilization. If there is to be unity signal at some signal level the stabilization must apply at all signal levels.

The use of inhibition alone will not permit decay of a 'flood' caused by an unusually large signal. Far more drastic action must be taken.

Inhibitory control is akin to negative feedforward in analog systems. However, it does have certain of the characteristics of 'delayed negative feedback control with very short recovery times'.

If a matrix is to include memory properties, 'real' or 'wanted' signal memory must be clearly separable from the apparent memory trace caused by a 'flood'. A memory trace must be set up by a repeated signal; a flood occasioned by a signal transient must be caused to attenuate rapidly. Better that the matrix be insensitive to any signal component that would be likely to cause a flood.

7.7 SHUTTERING

Amongst the most important of the mechanisms to be used in auto-stabilization is that of shuttering. For successful operation of these matrices, steady signals should not be applied as they tend to lead to flooding. Instead, signals should be applied in 'bursts'. Application of a signal burst will result in the launching of the signal into the matrix. Its behaviour there will depend upon the nature of the matrix and on its current state (resulting from previous activity). The signal burst may well decay without travelling any significant distance. However, it will have raised the general level of excitation of many elements in the vicinity of the signal.

Repetition of the signal burst could well cause considerable matrix activity and a cloud-image of the signal could commence to propagate. Given suitable inhibitory activity, a stable moving cloud can be set up. If the signal is removed following a few bursts, it is quite possible for the now-stable cloud to progress through the matrix to reach the far end and deliver its message.

This form of activity has been studied extensively using the computer simulator system which is described in Chapter 12.

The combination of inhibition and shuttering goes a long way to achieve a number of the required characteristics of signal processing we may need in practice. Inhibition provides a measure of gain control, shuttering minimizes flooding.

Shuttering however, introduces another effect commonly encountered in the design and application of digital computers for analog signal processing. We are now effectively 'sampling' the signal in time. One of the immediate bonus points is that transients are far less likely to be sensed; those that are sensed are likely to cause a minor disturbance in the matrix – and this disturbance will generally decay during its passage through the matrix. However, problems with aliasing of signal frequencies can still arise just as with analog–digital conversion systems.

7.8 SELECTIVITY AND DISCRIMINATION

Given a complex signal, that is one having a non-simple spacial distribution and varying in time, the characteristics of the matrix can be seen to play an important role in selecting 'preferred' parts of the signal and discriminating against other signal components.

Weak portions of a signal may be unable to set up the information clouds which should carry them through the matrix. Strong portions of the signal would excite the auto-stabilization mechanisms and be strongly propagated.

By 'strong' we mean a signal of large amplitude (causing rapid sensor element triggering) or of extended dimensions. The latter form will cause multiple excitation of groups of in-matrix elements and hence set up a cloud.

Weak signals would be those components of small amplitude and those of restricted dimensions.

Dynamic auto-stabilization, especially by inhibition, effectively compresses the internal dynamic range of external signals. The selectivity and discrimination of matrices are dependent upon the degree of compression provided by the auto-stabilization mechanisms. The computer simulations used extensively prior to the preparation of this text confirm these conclusions.

The reader will observe that, once past the detailed design parameters of the LSE, much of our discussion has been non-algebraic. The decision to depend heavily on empirical studies using computer simulation was taken during the early 1970s. Those statements regarding the behaviour of matrices, given in a positive manner stem directly from empirical studies and are not simply tongue-in-cheek observations. Where the text is less positive in its 'conclusions' then neither fully rigorous mathematical treatments nor empirical studies have been attempted.

7.9 RESOLUTION UPGRADE BY JITTER

Difficulties with shuttering can be alleviated to a considerable extent if, instead of switching the signal on-and-off, it is simply moved cyclicly about the sensor-plane. An enormous range of options arises with this concept – the signal could be 'swayed' or 'nutated', circularly-scanned or rotated over even quite a small region of the sensor-plane.

With such techniques, difficulties of aliasing would be alleviated as all the information relating to the signal would remain within the matrix. With careful design of the scanning system, it would appear possible that the signal could be very neatly processed in terms of both time and space integration or differentiation. Auto-stabilization would play its part in achieving good control of the dynamic range of the signal-handling capability.

7.10 DYNAMIC PARAMETER STABILIZATION

Forms of AGC can be envisaged such that, if elements are sensitive to changes in supply potentials and demand more current when activity increases, these increased power demands can be used to reduce the elemental sensitivity. We used to call it 'poor regulation' and have become accustomed to demanding very high levels of power supply regulation.

We now suggest a reversion to the older, simpler and more variable form of active processor. Whilst inhibition does provide auto-stabilization, its effects are applied to individual elements. The use of poorly-regulated multidrop power supplies would presumably provide some further degree of auto-stabilization – but covering regions of a matrix rather than the individual elemental control afforded by inhibition.

Simulations of effects of this nature have been used repeatedly in computer models of matrices – and the effects accord with our intuitive expectations.

7.11 RESPONSE TO NOISY SIGNALS

Noise is non-systematic in both time and space. The combined effects of selectivity and the poor response which may be expected with asynchronous effects should provide a fair degree of noise immunity. In fact the situation should be even better than we may expect. The processing abilities of pulsatory matrices extend to the discrimination against disturbances – even those which are 'carried' by real signals.

We may well expect that 'noise spikes' will create rapidly-decaying 'clouds', that both space- and time-integration of noisy signals will lead to clarification of signals and that the various mechanisms of auto-stabilization would rapidly restore normal response following a large burst of noise.

7.12 ENHANCEMENT OF LOCALIZED FEEDBACK (BACKCOUPLING)

A relatively simple modification to the $\boldsymbol{x}$ component of an interconnection specification can considerably affect the performance of a matrix. Instead of using: $0 < \boldsymbol{x} < \max(\boldsymbol{x})$, use: $-1 <= \boldsymbol{x} < \max(\boldsymbol{x})$. that is $\boldsymbol{x} := \boldsymbol{x} - 1$ or even $\boldsymbol{x} := \boldsymbol{x} - 2$.

Thus, a simple interconnection specification may be:

$\boldsymbol{x}$: = Poisson Distribution with $m = 3$

y: = Normal Distribution with $m = 0$, sigma = 1.

This could be modified to:

$\boldsymbol{x}$: = Poisson Distribution with $m = 1.5$

or $\boldsymbol{x}$: = -1 + Poisson Distribution with $m = 3$.

Then, a small proportion of the outputs from an element will feed to local elements in the preceding rather than the succeeding plane.

The effect is to introduce a form of localized feedback which will perform a small degree of image enhancement. This was the first of the 'response booster' techniques used with early computer models. Certainly there is an increase in the 'gain' of the matrix. With this comes increased discrimination against weaker signals in favour of strong signals. This is a standard option available in current computer models of pulsatory matrices.

7.13 THE EFFECT OF DEFECTIVE COMPONENTS

As the complexity of our matrices increases in terms of backcoupling, inhibitory control and auto-stabilization of elemental parameters, so the dependence of our matrices on individual elements reduces.

When using the full range of complex options available, it is readily demonstrated with computer models that individual elements have little effect on signal processing. Many simulations have been run using randomly-dispersed 'defective' elements with little observable effect on processing. The computer models used currently all include a complete row of defective elements. During demonstrations of the models, the existence of the defective elements has to be pointed out to observers and even then it is quite difficult to see the effect of a completely defective row.

8

Linear reflexive memory

8.1 GRAPHIC REPRESENTATION OF MATRIX INTERCONNECTIONS

Before launching into this next 'complication' in the development of pulsatory matrices, we shall pause for the introduction of that obvious graphic simplification of a three-dimensional network. Figure 8.1 provides the concept of a 'slice' from a matrix, just one element in thickness. Using this technique, we can examine plane figures for the derivation of new concepts and illustrate the effects of certain design parameters using plane computer graphical representation.

A representation of a simply-connected, that is, undirectional, plane is given in Fig. 8.2. Here an arrow is used to indicate the general trend of interconnections. One can read each arrow as a symbol for a complete pulsor element with all of its input and output lines, the general pattern of the interconnections being in the direction of the arrow.

This symbol may be taken to include the 'local feedback' form of connection, that is, that a small proportion of outputs are taken from two local elements of the preceding row (plane in the three-dimensional case). The symbol then, can be read as a normal Leaky Summator Element with the probabilistically-determined quantity of inputs. Such probabilistic 'wiring' details include, of course, intraplane connections – in the two-dimensional diagrams, these will be intrarow connections.

Initial investigations were conducted using matrices of forward-connected elements as indicated in Fig. 8.3. Such arrangements however, fail to meet a number of our performance and processing requirements.

8.2 REVERSED ELEMENTS

A 'reversed element' is one whose direction of interconnections along the principal propagative direction is reversed. If the space vectors x, y and z are used to determine the relative positions of elements receiving impulses from the given element, then a reversed element would be coupled to other elements at positions $-x$, y, z. Using our plane diagram technique this would reduce to $-x, y$.

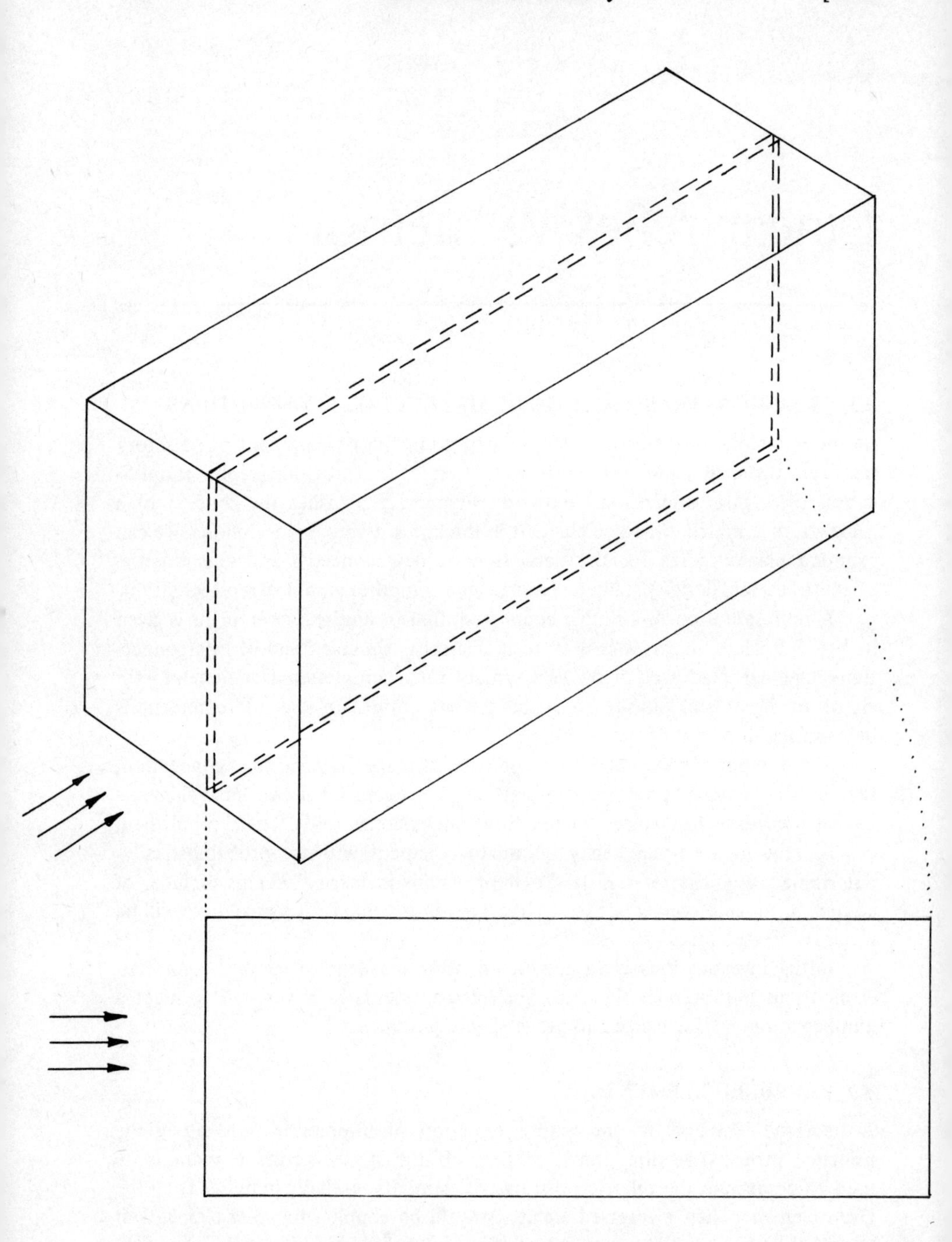

Fig. 8.1 – 3-D to 2-D Simplification Diagram.

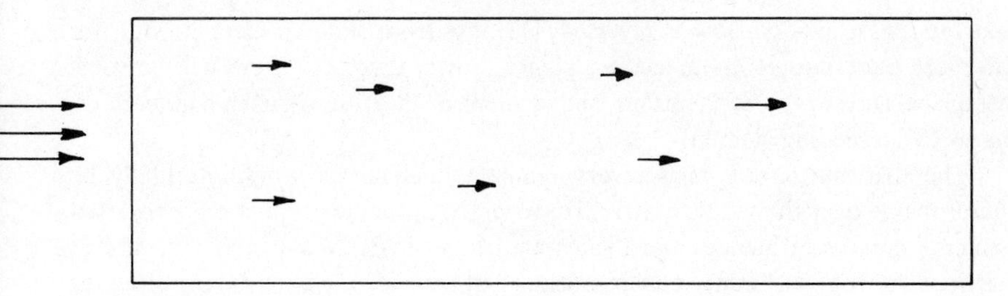

Fig. 8.2 – Signal Representation diagram.

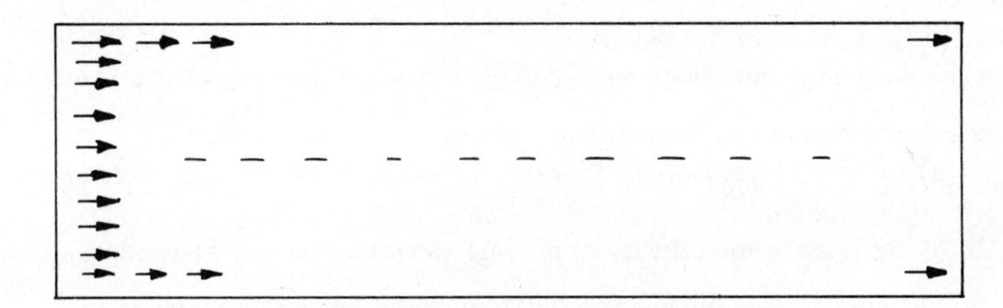

Fig. 8.3 – Plane of Forward-connected Elements.

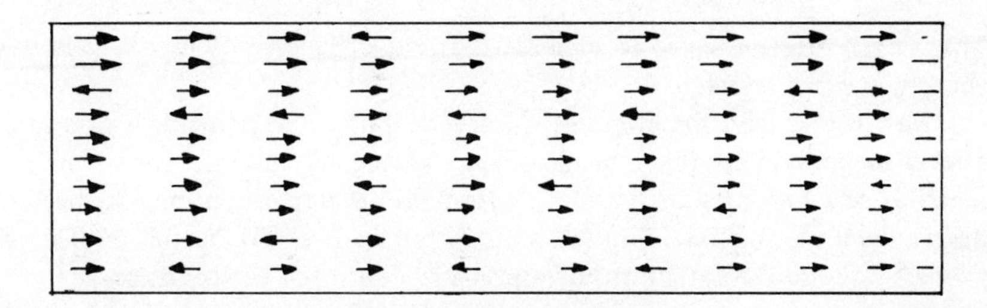

Fig. 8.4 – Use of Reversed Elements.

Reversed elements will of course tend to oppose the normal direction of propagation. They are thus a source of feedback. It is not right to deduce the sense of the feedback directly due to the effects of plane bypass which may introduce noticeable phasing effects. However, the feedback will tend to be 'positive' inasmuch as it will tend to reinforce signals. From this a number of characteristic performance patterns can be deduced.

Given the plane-with-arrows diagram, the definition of a reversed pulsor element is coupled with a reversed arrow. Figure 8.4 illustrates the concept, there being a small proportion of reversed arrows. A reversed element is connected using the same set of rules as apply to a forward-connected element save that the major axis of flow is reversed. Outputs from such an element will, for the most part, couple to preceding planes (rows) save that there will be some intraplane (intrarow) connections and a small proportion of back-connections, that is to succeeding elements.

The distribution of such reverse-connected elements would normally be equiprobable over the whole matrix. However, because the proportion of reversed elements greatly influences the behaviour of the matrix, one may well design matrices having different concentrations of reversed elements in different regions of the matrix. The purpose would be to secure specialized propagation and processing performance in an overall system.

8.2.1 Effects of Reversed Elements

Immediately apparent effects of a small proportion of reversed elements include:

(a) increase in matrix sensitivity to certain types of signal
(b) decrease in sensitivity to other types of signal, that is, increased discrimination
(c) decrease in the velocity of propagation under certain conditions.

Consequent upon the increased discrimination are such effects as:

(a) certain signals fade and fail to propagate
(b) certain signal conditions cause the body of the matrix to 'flood'.

The overall effect then is to increase the rate of change of matrix gain with changes in signal levels.

Not surprisingly, the effect of the reverse-connected elements is akin to the effects of positive feedback in the simpler analog systems. We must search for methods of achieving stability of some form. For preference, noting the need for high reliability, such stabilization as is introduced should be full auto-stabilization – even to the extent of fully optimized control of performance.

8.3 FACTORS AFFECTING MATRIX PROPAGATION

The characteristics deduced for LSEs in Chapter 4 form the start point for assessment of the performance of the more complicated forms of pulsor matrix. A summary of the major factors involved in matrix propagation is given in Table 8.1.

A large proportion of the research undertaken into the theory and design of pulsatory matrices has been devoted to the formulation of concepts and their description. At the present stage of development, it is quite unrealistic to provide actual equations. Because of the need for auto-stabilization, provided only that stability can be assured, detailed mathematical descriptions in the conventional algebraic forms would be somewhat of an overkill and help little in either further investigation or design.

Table 8.1
Factors Affecting Matrix Propagation

Factor	Effect			
	Sense		Type of Expression	
	Positive	Negative	Simple Algebraic	Exponential
Elemental sensitivity	X		X	
Trigger level		X	X	
Decay		X		X
Complexity (inputs per element)	X		X	
Bypass		X	X	
Spread		X	X	
Local Feedback	X		X	
Reversal	X			X
Inhibition		X		X
Signal status	X			X
Activity	X			X

Note – A pair of crosses on a line in the body of Table 8.1 indicates:
(a) the sense of the change in the propagation as the associated factor is increased;
(b) and the type of expression involved in that factor.

The most important tool in current use is the computer simulation described in Chapter 12. This enables exceptionally complex matrices to be studied dynamically. Use of the simulator is based on the principle that, analysis of a complex system is often best tackled by successive approximation using synthesis to isolate and evaluate the behaviour of the numerous factors involved.

The three illustrations Figs. 8.5, 8.6 and 8.7 indicate some of the propagation properties of a plane matrix. In each figure, there is an input on the left, covering the central region of the 'sensor' row. One of the possible requirements is that such a signal be transmitted through the matrix with minimum spacial distortion. Figure 8.5 indicates this desired state of affairs, the marks indicating LSEs which trigger in response to excitations.

When the gain locus is such that a signal is of insufficient amplitude to inject a self-sustaining cloud into the matrix, the situation is as indicated in Fig. 8.6.

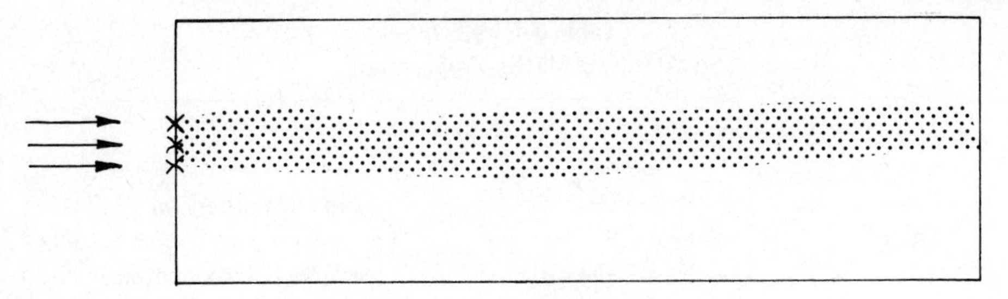

Fig. 8.5 – Desired Propagation Format.

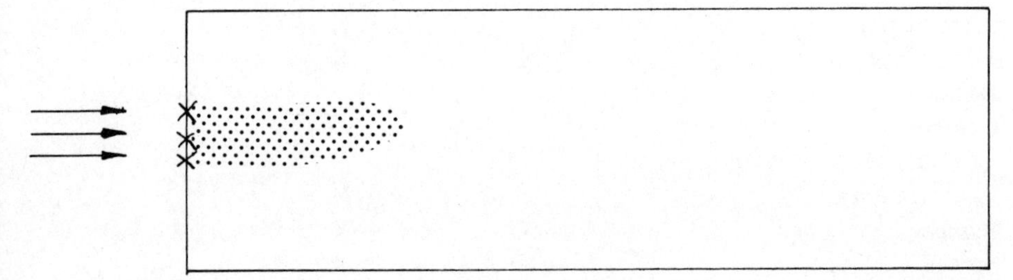

Fig. 8.6 – Signal Decay.

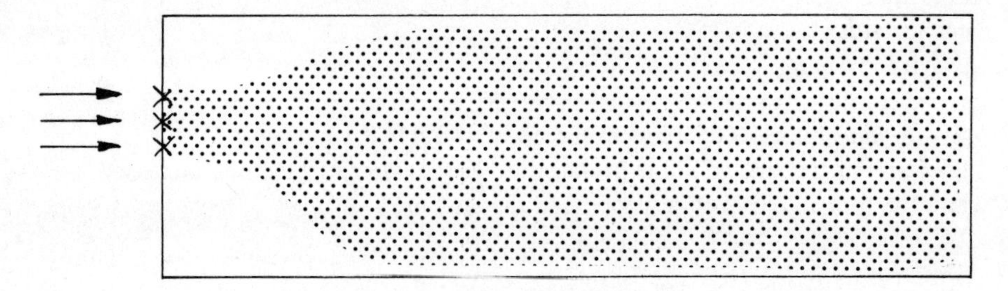

Fig. 8.7 – Matrix Flooding.

No actual cloud forms and hence the signal decays. The effect bears some similarity to that of a passing cloud – the local activity will cause many elements to receive inputs, but these will be insufficient to cause triggering in most of the elements. Were the input signal to be removed, there would be a 'seeded' region in which a number of elements are at higher than 'ground state' and hence are in a more receptive state than other (unseeded) elements.

When the critical stage of gain = unity is exceeded, then a general spread of activity occurs until the matrix floods as indicated in Fig. 8.7. Much of our work is in the prevention of this condition.

8.4 PROPORTION OF ELEMENT REVERSALS

Very low concentrations of reversed elements cause little real effect at all. There will be some minor increase in matrix gain and discrimination. Concentrations of a few percent elements do however, display very significant effects. The nature of these can be gleaned from the entries in Table 8.1. Very high concentrations of reversed elements generally produce highly undesirable increases in matrix sensitivity.

One effect which is quite pronounced is the reduction in the velocity of propagation of a stable, freely-moving cloud. Under certain circumstances, the regional gain can become so high due to feedback that a stationary cloud can be set up.

Given a stationary cloud, input signals will suffer considerable distortion which could give rise to some peculiar 'splash' effects. When a signal cloud impinges on a stationary cloud, the region of activity will spread horizontally. Given that outputs are taken from the periphery of a matrix adjacent to a stationary cloud, we have the elementary possibility of 'recognition' of the 'event'.

Figure 8.8 is included at this point to provide a graphic indication of a number of the effects discussed earlier. We take a plane matrix with backcoupling, element reversal (of a few percent) and inhibitory activity present. A shuttered signal is applied to the unseeded matrix. The illustration is of a series of snapshots of the state of activity within the matrix. Snapshot (a) indicates the initial application of the input which persists during periods (b) and (c) with increasing penetration of the activity. We assume that matrix and signal conditions are such that, when the signal is shut off at (d), the 'cloud' is detached and commences its journey. By time (e), the cloud is well on its way, leaving in its wake, the seed trail.

When, by time (f), the signal is restored, the original cloud is remote from the input region but the new signal-based activity propagates rapidly along the seeded region. The remaining snapshots indicate the slightly increased velocity of the new cloud. We can visualize that subsequent signal inputs of a similar nature will propagate at a velocity slightly greater than that of the original. There are obviously steady-state conditions for dynamic performance just as with the static gain calculations. We may well imagine a system in which signal shuttering is controlled by matrix activity.

8.5 AMPLIFIERS

8.5.1 Gain Factor

The gain of a matrix amplifier is dependent both upon the state of that part of the matrix in which activity is induced and upon the matrix parameters – complexity and sensitivity.

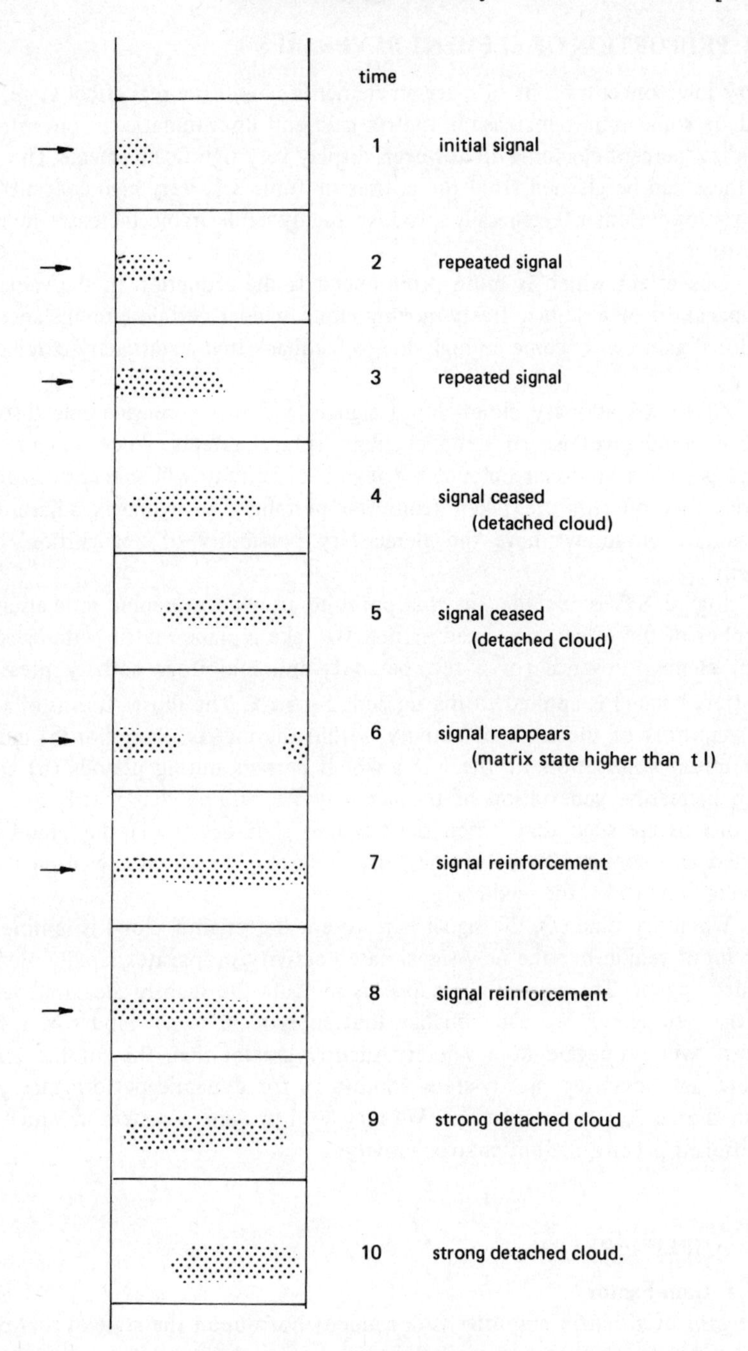

Fig. 8.8 – Series of Matrix States With Shuttered Signal.

The parametric gain is naturally limited to unity for strong signals. For signals below a certain critical level, gain falls to below unity. Between these low and high limiting signal values, there is integration in both space and time. As a result of these compound factors, gain can be very high. In a seeded region, a single pulse may cause the induction of a cloud. If a weak signal remains for a considerable period, activity will be induced. Shuttering can then commence and the 'weak' signal can be handled as readily as a strong signal which receives a lesser duty cycle for initiation.

8.5.2 Bandwidth

The initiation of a cloud into an unseeded matrix region has been considered in some detail. The time taken for formation of a cloud will be of the order of a few 'glimpses' of the signals via a shuttering mechanism.

Propagation velocity is initially low – a few planes per glimpse. Once seeding has occurred, velocity may reach a few planes per pulse period in response to a 10% duty cycle.

The decay of a cloud on removal of its initiating signal depends primarily upon matrix parameters together with the log-dec. of the elements composing the matrix. A signal cannot be extracted from a matrix once actual pulsation has ceased – even though the seed remains.

At a given pickoff region, activity will be the result of a transitory cloud. Decay then, is dependent upon the effective length of a cloud. Typically this can be from a few times the plane bypass 'length' to the length of the matrix. In the worst case (a full-length cloud seed). Transit time for the tail of the cloud is parametrically dependent and can be a few shuttering periods.

8.5.3 Signal Enhancement by Jitter

Given a signal that is complex in both time and space, we question the space-time resolution of the signal by a pulsatory sensor: processor matrix. Dynamic resolution is limited by various factors – the most obvious being the number of sensor elements in the sensor plane. Unfortunately, unlike other sensory systems, the pulsor system normally requires to be presented with components of the signal on a number of sensor elements. Such 'diffuse' sensing offers the advantage of high gain and hence high sensitivity. The pulsor device then appears to play off sensitivity against resolution both in space and time.

However, for a signal to contain 'useful' information, there must be 'structure' of some sort in the signal as presented to the sensor plane. The sensor: processor device will contain some shuttering mechanism. During a period when a given portion of the information field is cut off from the sensor, a different portion of the field could be applied to the sensor. A degree of seeding will be afforded to the whole sensor plane and matrix and hence the matrix response to incoming information will be quite high. There could be some confusion in time as the information field is swept over the sensor. However, we shall demonstrate mathematically in a later work, how this modulation effect can be turned to advantage in such a way as to actually increase the potential resolution of a pulsor system whilst retaining relatively high sensitivity.

8.6 THE PROCESSES OF FLUSHING

A matrix may be in a number of generalized states. It could be:

(1) processing normally
(2) reaching saturation in terms of information packing
(3) deactivating due to lack of input
(4) inactive with low-level trace seeding.

Given a 'swamping' or 'flushing' input (an input spread equally over all sensors), the matrix can react in one of a number of modes:

(a) saturate and destroy existing information
(b) experience an increase in sensitivity and hence enhance activity
(c) reactivate existing seeded trails.

By careful selection of the purpose of and intensity of a flushing signal, it becomes possible to design further control mechanisms. An overactive state in which information units were merged rather than simply propagated or correlated could be controlled. The matrix would be over-excited and excessive inhibition introduced. The result could be a sensitive matrix, highly susceptable to new inputs, and with fast propagation times due to the generalized seeding which would result. Again, a flushing signal could be used simply to enhance the sensitivity of a matrix. Yet another possibility is that, when activity ceases and a seeded trail is all that remains, a simple flushing action would restore activity – a form of memory recall.

8.7 PROPAGATION VELOCITY IN A PULSOR MATRIX

Propagation velocity is highly dependent upon the state of activity or 'seeding' of the matrix. If a signal has recently moved along a portion of a matrix and, following a shuttered period, that signal is temporarily re-instated, then velocity of propagation will be remarkably high. Effectively a 'track' is formed along the matrix and the renewed signal can follow the track very readily.

Where the state of activation of a matrix region is very low (no activity for some considerable time) then a number of signal 'flashes' will be required to set up a 'track'. Reasoning on from this point rapidly discloses that a seeded matrix will be exceedingly selective in favour of the signal which caused the seeding and highly discriminative against other signals.

Simulations using some hundreds of planes of 'LSEs' have very clearly exhibited these effects. The combined use of backcoupling, reversed elements, inhibition and high complexity in simulated matrices has demonstrated a remarkable degree of fidelity in 'cleaned-up' signals reaching the end of long matrices. The problems of stabilizing signals during propagation are well on the way to satisfactory solutions.

8.8 VELOCITY MODIFICATION BY ELEMENTAL REVERSAL

8.8.1 Initial Velocity

The drift velocity of the tail of a cloud is retarded in the presence of reversed elements. This is readily appreciated when considering that, during a shuttering period, elements closest to the sensor region will continue to receive some inputs from the triggering of reversed elements ahead. The tip of the cloud will penetrate more rapidly in the presence of reversed elements. This is due to signal enhancement which tends to maintain activity during signal shuttering.

The result of these two effects is that clouds are extended in the direction of propagation when reversed elements are used. The stability in terms of activity limitation is ensured by the use of correct levels of inhibition.

8.8.2 Velocity in a Seeded Matrix

Once a cloud has been established and has commenced its journey through the matrix, it leaves a 'trail' of partially-activated elements. This trail will decay logarithmically at a relatively low rate (determined by the log. dec. of the elements. Any new activity such as the reappearance of the original input will find a ready path for propagation. The velocity of a cloud in a seeded region of a matrix is therefore relatively high.

8.8.3 Discrimination by a Seeded Matrix

Weak or transient signals cause little activity and hence result in weak trails. The natural discrimination of a pulsor matrix in favour of a strong signal is enhanced when a signal matches its seed trail. The combination of these effects when a repeated (shuttered) signal is handled increases the signal contrast, minimizes the weaker components and provides a considerable degree of protection against unwanted transients present in the signal.

8.9 THE INTERPLAY OF PARAMETERS

During simulations of pulsatory matrices, initial signal duty cycles of 0.4 have frequently been used to initiate propagation. Once a cloud has been set up its resulting seed effect permits the duty cycle to be lowered to below 0.1 whilst retaining high-velocity propagation. For one experiment, a cloud was induced and its stimulus removed. The cloud activity fell off as propagation carried the main body of the cloud away. The remanent seeding was allowed to decay very considerably and then, the sensitivity of the matrix elements was artificially increased. Activity recommenced, an auto-stabilization algorithm reduced the sensitivity once more and the revived cloud propagated the length of the matrix.

From this experiment it may be seen that there are many effects which can interplay in the pulsatory matrix. Given sufficient complexity to encourage activity, a matrix can be made to operate stably under wide-ranging conditions. Furthermore, the presence or occurrence of defective elements will have a negligible effect until the proportion of defective elements reduces the operational complexity of the matrix to an unacceptably low value. Commencing then with an abundant complexity value, very considerable degradation can be tolerated and even go unnoticed because of the self-adjusting nature of the matrices.

8.10 INTERACTIONS WITH EXTERNAL FIELDS

8.10.1 Fields produced by pulsor matrices

The lack of synchronism in the triggering of elements means that there will be considerable integration of the electrical fields produced by a 'plate' of elements at a distance of a centimeter or so. Such fields as are produced will be predominantly electric, unipolar but with considerable noise present. Given a matrix of centimetric dimensions, then at a short distance, there will be an electric field due to the summation of all the pulsor activities. We could assume that activity occurs using some 10–30% of elements at any one time. This would represent the handling of all the signals and the activity of all the clouds. The integrand of the resulting electric fields would tend to constancy at a few centimetres from the matrix.

The task of decoding the pulsor activities by examination of the external fields then is one of poor returns. There being no underlying synchronism, there are no base references to work from even if the fields could be sensed. Some of the memory operations develop a form of synchronism which may appear as more or less regular fluctuations but containing no discernable detail.

Significant magnetic field components may be expected only from the operation of the power supply for the matrices. However, the integrand of current fluctuations will follow the same sort of pattern as will the external electric field. The information content of the external magnetic field will be exceedingly small.

The combination of the electric and magnetic vectors associated with electrical activity can, under certain circumstances, produce a 'radiant' field. This occurs when (a) the two fields are in time phase, (b) are in space quadrature about a common origin and (c) have the fixed ratio necessary for the launching of a radiant wave. The power produced in such a wave is dependent upon the dimensions of the source and the frequency of the synchronous fields.

Because we have none of the conditions required for the set-up of an appreciable pair of synchronous field vectors, the risk of transmitting useful information from a matrix is very low.

8.10.2 Sensitivity to External Fields

The question here is 'What is the field condition which could affect the behaviour of a pulsor matrix?'. Consider firstly the effects of external electric fields. An external d.c. field will affect all parts of the matrix virtually equally. Some remanent component may exist along the planar axes of a slice. This will however, be of such small magnitude as to have virtually no effect other than perhaps a minor effect on matrix sensitivity. The self-regulating quality of the matrix will provide protection against this.

Protection against very large electric fields may be achieved by use of any of the well-known techniques of screening the field equalization.

The effects of an external a.c. field may readily be deduced by noting that there will be effects significantly different from those at d.c. only when the external field frequency is high. Again, the classical screening and shielding techniques will produce devices insensitive to such effects.

Similar reasoning can be applied to the effects of external magnetic and radiant fields.

The pulsor matrix then, can be seen to be relatively immune to either annunciation or response to unwanted stimuli.

8.10.3 Effects of Radiation and Particles

The effects of radiation and particles on the hardware of the pulsor system would be in essence, no different than are the effects on other microelectronic equipment. There is a small risk that small sections of a circuit could suffer damage. However, this will not affect the operation of the system for reasons discussed elsewhere in the book. The effect of such attacks upon the data held in LSEs would be negligible – again for the same reasons.

This contrasts with the effects of particles on digitally stored information in microelectronic devices. If a single bit is altered in a computer system, the effects can be devastating in any system not equipped with error detection and correction hardware. When a stored bit is affected without protection, the effect may not be obvious for some time – until that bit is referenced. The result could be anything from a trivial change in the value of an item of data to a catastrophic program flyaway.

8.11 LIGHT GUIDES IN PULSOR PROCESSORS

One of the requirements of a sensor is that it must be capable of collecting information from some source and passing it to the sensor of the processing matrix. In a number of cases, the sensor will be optical in nature. Scanning will frequently be a mode of use, for example to obtain resolution enhancement or shuttering. The scanner should operate on a lightweight structure with the appropriate degrees of freedom.

An obvious way to isolate the matrix from the optical assembly mechanically is to use mirrors or light guides. The former will be omitted from the discussion for obvious reasons and attention will be given to the use of light guides. Again, the obvious form to consider is the fibroptic bundle.

The 'coding' in a bundle of fibroptic light guides is rarely simple. By this, we imply that rarely will you find a bundle in which each end of each fibre holds corresponding positions in the face geometry of the bundle. For some communications systems this has been a source of considerable embarrassement in design. For the pulsor matrix system, this should give little cause for concern.

The puslor matrix relies on there being defective elements and the mispositioning of one end of a fibre would give rise to effects similar to noise. Many forms of noise are processed out by a matrix. The overall effect then of misplaced fibres is simply a slight diminution in sensitivity.

8.12 THE POWER OF COMPUTERS

Because the computer is the principal agent in the pulsor research, we shall spend a moment considering the true effective power of a number of classes of computer. The spectrum of computers will be split into four groups: microcomputers, mainframes, supercomputers and distributed computer systems. The measures used will be the MIPs (million instructions executed per second), the MFLOPS (million floating-point operations per second) and the MIOTS (million input–output transfers per second). Typical current figures for these are given in Table 8.2.

Table 8.2
Speeds of computer systems

Class	MIPS	MFLOPS	MIOTS
microcomputer	.1–1	.001–.1	.001–1
mainframe	1–10	.1–1	.1–1
supercomputer	10–100	1–50	.1–1
distributed processor for example 100 micros + supercomputer	100–200	1–100	1–10

The fact that it is obscured by these (quite impressive) figures is that a calculation normally involves many 'machine cycles' – possibly many 'instructions' and 'FLOPS' so, to perform relatively simple tasks such as convert some incoming data from polar coordinate representational form to rectangular to suit a particular plotting device may involve hundreds of individual calculations. The throughput time for such work then is a more sensible figure to use and we enter the rather contentious world of 'benchmarking' the performance of computers. A crucial factor in the design of computer programs is the mapping of the task onto the architecture of the available computer. A feature of many 'super-computers' is that their high performance is dependent heavily on careful problem analysis and hardware mapping. Should you wish to process a 64 X 64 matrix on a particular computer, then the internal processing time may be a microsecond. For a 65 X 65 array, the processing time may well be minutes.

Very considerable differences are evidenced by the sequences of computer code produced by the translators (compilers, interpreters, etc.) associated with individual computer programming languages. Even as this manuscript was being prepared, a report was received citing a PASCAL compiler for a particular computer which produced code taking some 500 times as long in execution as that from a FORTRAN compiler on the same task and computer.

The fact that computers can be used to perform data processing at all should be considered as one of the remarkable achievements of the 20th century.

Special-purpose computers have been constructed to perform simple processing of TV signals in real time. The operations are simple correlations between adjacent elements on and between line scans. No ordinary computer is within orders of speed magnitude of the requirements for this task.

Associative field processing (such as the clean-up TV pictures and industrial pattern recognition) is a task for which the computer is but poorly suited. The Distributed Array Processor handles bit-mapped arithmetic on a 64 X 64 element array, that is, numeric 'words' are treated in bit-serial form to provide for arithmetic operations between adjacent elements of the array. The Sonar Image Processor of Chapter 1 uses a matrix of a hundred computers in three dimensions to perform essentially simple processing and provide a data-reduced graphical display in real time. Some hundred or so computers operate synchronously yet the graphical field holds only a few hundred picture elements.

The PULSENET simulator (Chapter 12) operates in a very fast computer yet can handle only a dozen or two triggering elements per second.

A wide gap exists between the capabilities of digital computers and the needs of realistic, real-time associative field processing.

9

Transverse memory

9.1 THE TRANSVERSE MEMORY TECHNIQUE

The transverse memory technique is based on the linear reflexive memory from a functional point of view but its driving and readout machanisms differ considerably. The development of this class of device will lead to a dramatic alteration in our concepts of system design.

Firstly, take a linear reflexive memory matrix as a fundamental component whose characteristics are now understood. Wrap the matrix end to end so as to produce a circulatory device (Fig. 9.1). Patterns induced into the circulatory matrix will have to take on a synchronous behaviour unless the matrix is very long.

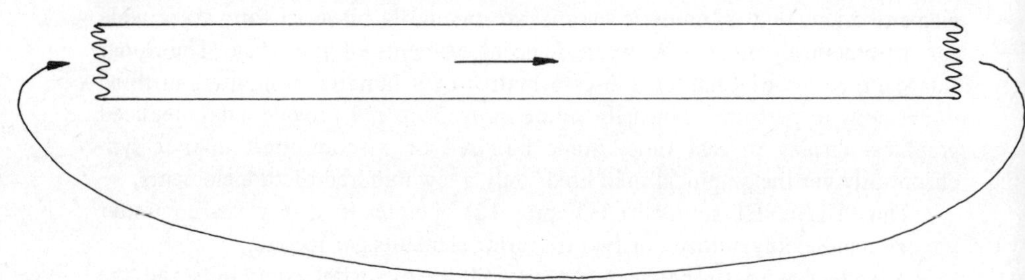

Fig. 9.1 – A Cyclic Memory.

A given pattern will circulate at a velocity dependent, as always, on the nature of the signal. The repetition frequency is dependent also upon the matrix dimensions. Given a sufficiently long matrix, the remanent seeding will decay behind the active cloud so that the cloud normally meets virtually inactive elements.

However, the matrix may carry a number of time-spaced clouds which do provide seeded trails and there can be some intermodulation of the clouds. This implies that each cloud can carry remanent information belonging to other circulatory clouds and a further form of diffusion develops.

Resulting from this, there is a further 'softening' of the signals carried in the memory system which can be disadvantageous but can also offer some interesting system behavioural characteristics which we shall exploit later. Basically, this mutual interference can be a distinct aid in active pattern recognition.

Figure 9.2 illustrates one of the many possible ways in which a sufficiently extensive ring memory may be fabricated within a limited space. It is one of the many possible configurations for microelectronic fabrication on silicon.

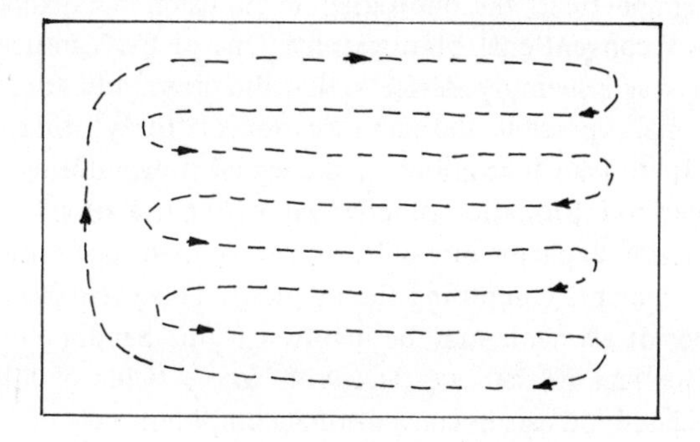

Fig. 9.2 – A Folded Memory.

9.2 MEMORY CAPACITY

A number of difficulties arise in the assessment of this parameter. The memory systems is not digital, neither is it analog. Just as with the pulsorcube theory, we find that almost antiintuitively, the capacity of the large pulsor matrix is not dependent upon the exact number of its elements. Further, if we consider that the purpose of the pulsor elements is 'to support patterns', then we can dissociate the elements from the patterns of activity which they support.

Now comes a freedom of flexibility which may come as a surprise to the thinking of the conventionally-trained electronic designer. The theory and practice of the holograph bears some relationship with that of the pulsatory memory system. The information is in the form of clouds of activity, mostly moving in circulatory paths about the matrix. Destruction of individual elements does not affect the supported patterns which are maintained by the complex activity of groups of elements. Propagation speed can be quite high, the density of clouds can also be high. We play off space against time, using multiprocessing on a scale quite impossible with digital computers.

The seeded trail associated with every activity cloud will tend to cause intermodulation effects between the clouds in a stream. Further associative effects will be found between clouds following 'adjacent tracks' through a matrix. The result is that the stored information relates not so much to individual information units as to rather larger 'information packets'. The nature and form of these packets is an analog transformation of the external set of information.

There is a limit to resolution both in space and time within the matrix but very considerable difficulties arise in attempting to denote the memory capacity numerically. One can draw a rough parallel with the different styles of usage between analog and digital recordings on magnetic surfaces. The packing denisty of a videotape recording greatly exceeds that of digital recording even though the same medium is used to the limits of its capacity in both systems. Again, the holograph treats the photographic emulsion in a manner quite different from that of conventional photography. One of the significant features of the pulsor matrix as a memory device is that the intermodulation effects within the 'recording' make possible the identification of 'fuzzy', that is, incomplete, sets of information. Transformations of the signal stream during 'recording' will equally be applied to information supplied later for correlation.

The information capacity of a pulsor memory then can considerably exceed the number of elements composing the memory. This even though a very considerable number of elements may be involved in the handling of each atom of information. The fact is that we deal with whole fields of information, not isolated or associated 'bits' as in conventional computers.

For very small matrices, memory capacity is readily seen to be rather less than could be supported by the same number of transistors using conventional techniques. With matrices having many thousands of elements, comes a break-even point in that calculated capacity approximates to the number of elements. But for memories holding tens of millions of elements, real advantages accrue in that a vast increase in storage capacity is achieved. The number of patterns supportable by the matrices becomes very large. The fact that all details are fuzzy but that the environment can be scanned aids greatly in the exploitation of the memory capacity.

There comes a point at which philosophical argument commences as to whether the capacity in terms of numbers of patterns is a denumerable quantity or whether it can actually become transfinite. The Author has enjoyed many discussions and arguments of an intensely philosophical nature over the past twenty years but has reached no conclusion on this topic. Large matrices have exceptionally large memory capacity.

9.3 THE CROSS-SECTION OF A CIRCULATORY MEMORY

It would seem reasonable to construct a memory channel from a number of 'bundles' of narrow cross-section rather than from a 'solid' matrix of LSEs. Some possible bundling structures were considered in Chapter 5. In micro-electronic fabrication we may well be constrained to the use of planar silicon structures for the forseeable future. It may therefore be necessary to consider plane sections along matrices. Such arrangements may well constrain our designs to the use of plane memory systems in which relatively narrow channels of LSEs are used as our bundles. A convenient mechanism for the segregation of such bundles could be the power gridding required for operation of the electronic devices. We could well consider that memory devices are to contain long and convoluted channels whose width is a few tens of LSEs.

9.4 MEMORY COUPLING TECHNIQUES

9.4.1 A Transverse Injector

Some interesting possibilities emerge when considering a junction between a unidirectional propagative matrix and a ring memory. If the junction is formed between the output plane of the propagative matrix and an edge or a cross-section of the ring, then clouds could be launched in the ring under suitable conditions of synchronization. A shuttering action in the initial propagative path, synchronized to enable powerful signal injection into the ring could feasibly induce memory. Figure 9.3 illustrates this concept.

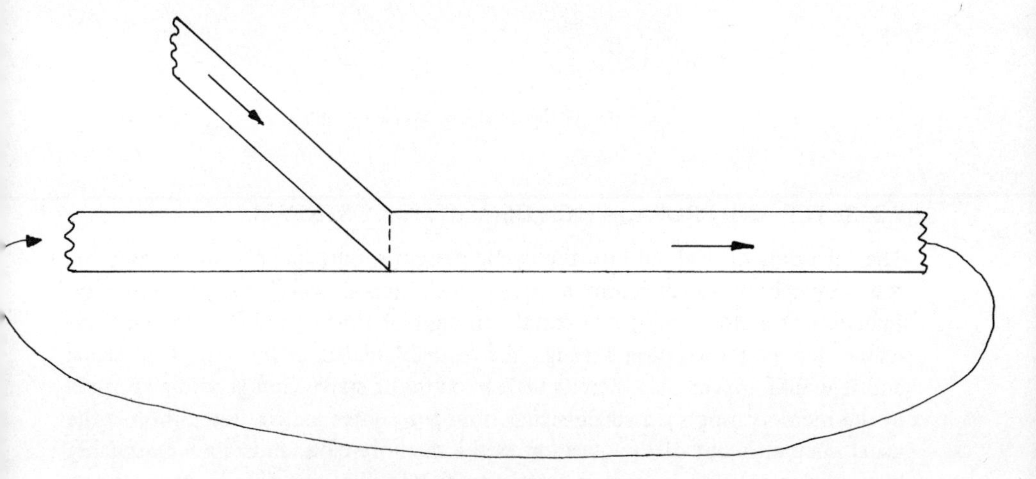

Fig. 9.3 – A Tranverse Injector.

Given also, a junction between the ring and the input of another propagative matrix near to the first junction, we could arrange that little or no coupling exists normally between the two propagative matrices. However, given that memory clouds are in circulation around the ring, a matching pattern in the 'sender' matrix could via the passing memory cloud, induce a propagative action in the 'receiver' matrix.

We then have a correlator – a memory retrieval embrio. Sets of such combination devices can be conceptualized with quite interesting possibilities opening up.

9.4.2 Orthogonal Transverse Coupling

Figure 9.4 illustrates a rather different form of transverse matrix coupling. Here, there is a translation between the transverse (space) coordinate of the input matrix and the longitudinal (time) axis of the cyclic memory matrix. This can lead to a rather different form of correlation and a different form of correlator output signal. This form of coupling will be investigated in some detail in Chapter 10.

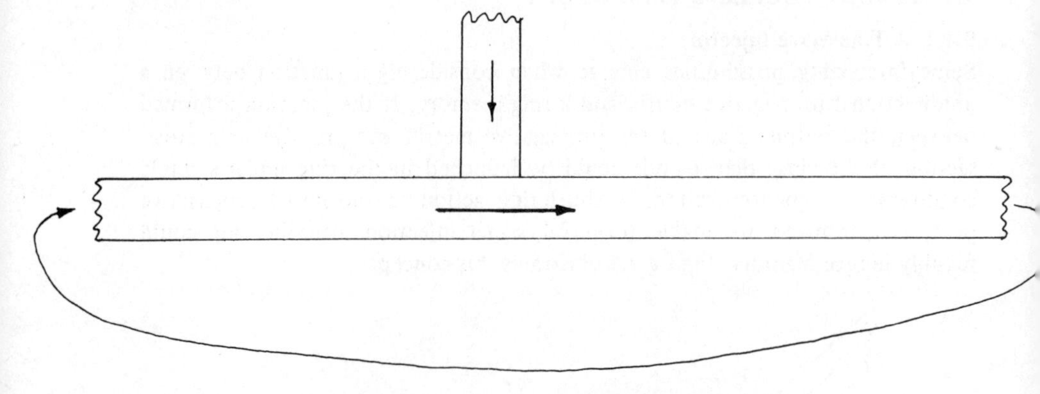

Fig. 9.4 – Orthogonal Transverse Coupling.

9.5 SETUP AND PROPAGATION OF A MEMORY STREAM

The injection of a cloud into the cyclic memory must be an activity requiring some degree of synchronism between the external world and the memory. Injection of a cloud requires a certain amount of time – and the cloud is a dynamic device. Propagation through the 'source' matrix is by 'bursts' of signal and it would appear that there is little hope that a stable cloud could be formed in the memory ring by a single signal from the source matrix. Repetition in the usual manner is out of the question as the memory cloud must have reasonably high resolution, that is, have a short time extent. We thus wait for the cloud to circulate the matrix and then arrange that, at the time the returned cloud arrives, so does a new version of the signal cloud. However, the length of the circulatory path is quite considerable and so, the initial appearance of the signal cloud is unlikely to trigger a stable pattern in the ring.

There could be a number of ways to circumvent this difficulty. One is to use a flushing input to raise the general sensitivity level of the ring so that a signal may be launched readily by a single appearance of a cloud at the injection point on the ring. This may prove inconvenient if the ring already contains information. A second method would be to arrange for a region of the input matrix to run parallel to the ring so that sufficient coupling exists for long enough to ensure a satisfactory signal launch.

9.6 TESTS ON CIRCULATORY MEMORIES

Some work was undertaken on this topic in the early days of discrete elemental fabrication. Some of the concepts of memory capacity were established during that period. CRT displays of memory packets indicated that very large numbers of distinct and stable patterns could be set up and supported using a single ring of elements.

More recently, computer simulations of planes containing many thousands of LSEs have been used. To provide a reasonable width of memory channel (say 20–30 LSEs), one is restricted to channel lengths of a few hundreds of LSEs only. Further, simulation time is such as to restrict the duration of an experiment to only a few thousands of elemental pulse periods. This is scarcely sufficient to ensure stability of a pattern. Much more work is needed in this very important aspect of pulsor technology.

10

Re-entrancy

10.1 TYPES OF RE-ENTRANT MATRIX

Two forms of re-entrancy will be considered here:

(1) Ladder networks.
(2) Reflective networks.

When these devices are considered as system components, we shall find a number of uses for combinations of these fundamental types.

10.2 LADDER NETWORKS

Figure 10.1 illustrates the concept of the ladder network. A pair of propagative matrices are set edge-adjacent with opposing directionality. Linkage networks are set between the 'input' and the 'output' matrices.

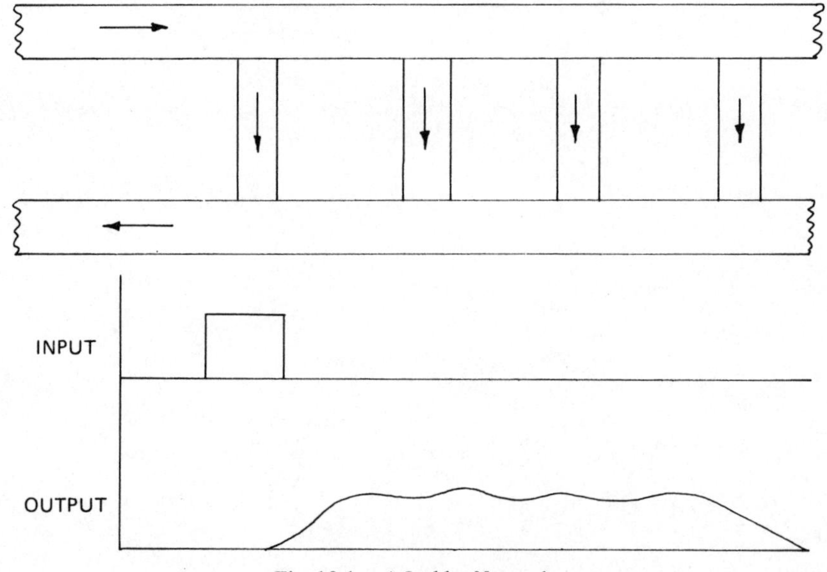

Fig. 10.1 – A Ladder Network

A number of pathways of differing length are now available between the sensor and the output regions. Consideration of the nature of response from this network shows that the output will commence fairly soon after application of a signal pattern as activity is initiated across the first rung of the ladder. Assuming that the signal progresses the length of the input matrix, then the output will carry an integral of the signal.

A 'weak' signal applied to the input of a simple ladder could produce an output response via the shortest route. Given suitable parametric design of the ladder, the additional pathlength via other rungs of the ladder may result in a decayed cloud and no further output. However, due to the seeding of the first signal burst, a further similar signal may well propagate further, resulting in enhanced output with further delay and 'output stretching'.

There are many potential applications for such behavioural characteristics in the processor elements of control systems. The matrix network provides powerful time integration features and the forms of gain optimization which can be designed into pulsor matrices.

By use of sets of such ladder networks, we can visualize new forms of system controller which include a variety of short-term memory features applicable to self-optimization.

10.2.1 Bi-directional Ladders

The bi-directional form of coupling illustrated in Fig. 10.2 will provide a further enhancement to the signal processing capabilities of the ladder network. Circulatory pathways can now be set up given suitable signal conditions and hence a new memory system begins to take shape.

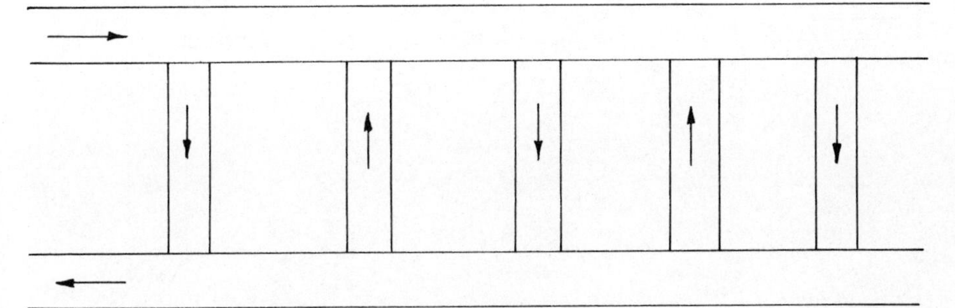

Fig. 10.2 – Ladder with Bi-directional Coupling.

Signals entering the matrix can set up cloud patterns with the possibility of forming looped sections. Signals able to cause such stationary loops will have set up highly active regions which can cause enhanced response to future copies of such signals.

Having a mechanism, we can design such networks so as to exploit their special characteristics.

10.2.2 Ladders with Transverse Memories

The next natural step to take would seem to be the introduction of a transverse memory system into a ladder network (Fig. 10.3). In addition to the normal characteristics of the ladder network, we would then have the response from a circulatory memory system. Here, we introduce the concept of synchronous response to certain classes of events in the external system. There would be the initial 'weak' response, the natural event-stretcher response of the ladder and finally an enhanced response due to the action of the memory. The latter would occur only when certain conditions of synchronization were met.

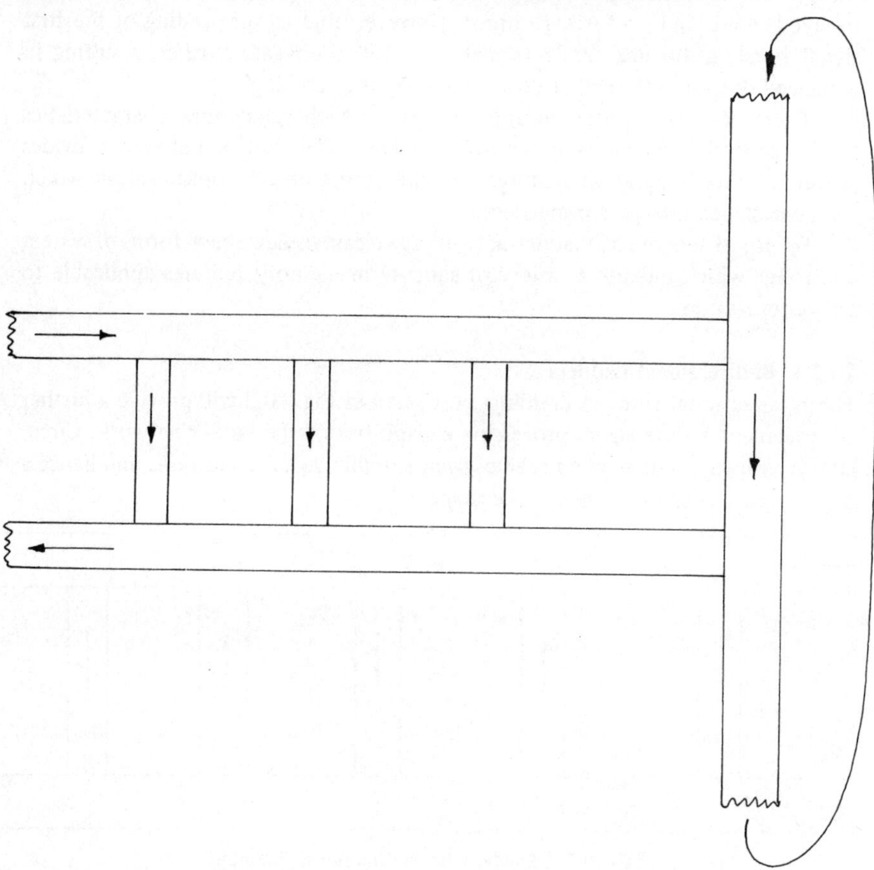

Fig. 10.3 – Ladder with Transverse Memory.

With the use of a set of ladders abutting the cyclic memory system we can see the development of exceptionally flexible system controllers. The set of ladders introduces a new dimension in parallelism and the whole is linked by the synchronous activity of the circulatory memory. Figure 10.4 illustrates the concept.

Fig. 10.4 – Set of Ladders with Transverse Memory.

10.3 REFLECTIVE NETWORKS

Consider the partitioned matrix of Fig. 10.5. Here there are virtually continuous couplings between the forward and the reverse flow sections of the matrix. By making the natural directions of propagation slightly skewed in relation to the matrix axes, we see another form of response arising.

There are local pathways from sensor to output surfaces, extended pathways and finally, a memory system coupling device. With this arrangement we blur the rungs of the ladders considered earlier and find a continuous development of pathlength according to signal and matrix conditions.

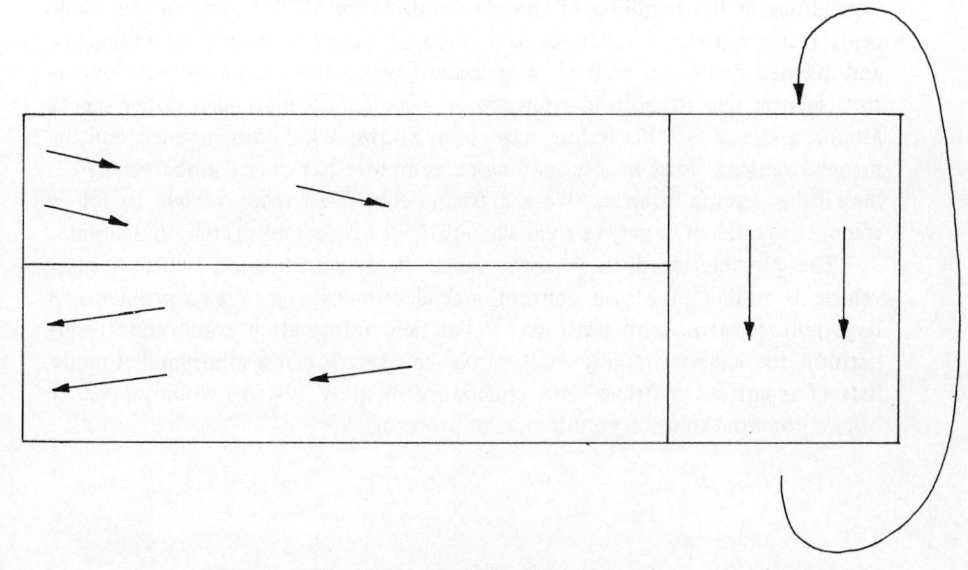

Fig. 10.5 – A Partitioned Matrix.

There are, in this arrangement, graduations of pathlength across the input and output edges of the system. Not only can signal strength affect the pathlength, the response is time-dependent upon the position of a signal on the sensor. This arrangement provides yet another signal transformation – signal position versus time. It is quite possible that some portions of the matrix could have propagation directions parallel with the matrix axes such that a variety of signal parameters can be assessed by the matrix.

A natural extension to this idea is illustrated in Fig. 10.6 which combines a number of the concepts discussed earlier.

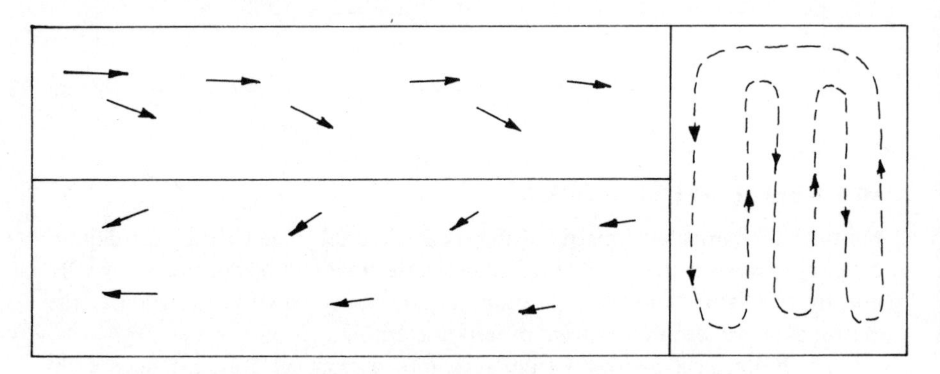

Fig. 10.6 – A Partitioned Matrix with Transverse Memory.

Multilayered structures of this nature would offer very powerful processing capabilities in the handling of complex information fields. Some of these concepts resulted from detailed consideration of the requirements of unmanned and manned spacecraft control. A spacecraft which loses some measure of control, such as due to radio interference or space debris, may need rather special attention. Cases of this nature have been known with both manned and unmanned vehicles. Loss of a ground-based computer has caused embarassment as have other events in space. Were a ground-controlled space vehicle to roll in response to a defect, it may be somewhat difficult to re-establish communication.

The possible need to perform stellar field identification whilst a space vehicle is rolling may give conventional electronics one or two problems. A large pulsor matrix with prestored stellar field information could conceivably perform the analysis rapidly as it would be operating in a multiparallel mode. Sets of re-entrant matrices with circulatory memory systems would appear to offer a potential solution to this class of problem.

11

Pattern recognition

11.1 THE PROCESSES OF CORRELATION

In statistical work with time series and in a number of aspects of signal processing there arises the need to compare pairs of data streams in order to derive a measure of similarity. The simple statistical test provides a single number – the correlation coefficient. For time series and signal processing work, there is often a need to derive such a figure at all positions along a data stream so as to provide a correlogram. In good cases, the corellogram provides a graphic indication of the patterns of similarity measure between the data streams. Such work imposes very heavy workloads on conventional digital computers and one frequently finds special purpose computing installations to deal with particular correlation problems. Such systems often use the techniques of **fast Fourier transforms** (FFTs) as a means of minimizing the amount of computation required by direct methods.

11.2 CROSS CORRELATION AND AUTOCORRELATION

The process of comparison between two data streams is commonly referred to as **cross correlation** (correlation across two streams). Reasons for using this process include the need to search a new data stream for any tendencies to relate with signal components in existing data streams.

This can be likened to an elementary form of memory retrieval. Processes of this type will find wide application in the field of pulsor matrices and networks. It will be appreciated that a signal component may occur at any phase relationship with a component stored in a circulatory memory. It is for this reason that so much computation is required to perform a cross correlation. We may foresee that an incoming signal may need to be correlated with signals in a number of points in the circulatory memory systems which have been discussed.

The second form of correlation we shall pause to consider is autocorrelation. In this, a signal is compared with a delayed version of itself. Any cyclic events in the signal could be mapped onto each other to result in a detection of the condition. However, to relate events of varying wavelengths requires that all portions of a signal are compared with many other portions of the signal as each 'length' of a correlation segment will indicate only certain wavelengths. Again, we may visualize a number of possible designs for pulsor matrices which could exhibit this characteristic.

11.3 PATTERN RECOGNITION PROCESSES

It is usual to consider that signal processing refers to single-channel signals (commonly excluding television) whilst the term pattern recognition refers to 2-D or 3-D information fields. However, in practice, scanning techniques such as a TV system are frequently used in pattern recognition systems. We shall permit ourselves the luxury of considering that our pulsor matrices will be capable of accepting a full 2-D image. Given a sufficiently large number of LSEs in a system, it is certainly feasible that an associative memory for 2-D images can be developed. Then it is a short step to the development of a full 2-D (even n-D) correlator which may be used in recognition systems.

In modern pattern recognition systems, a number of stages are used in the extraction of image data. One of the significant terms in this work is **Image Processing.**

11.4 IMAGE PROCESSING

This chapter can contain but a few brief notes on this very complex subject. However, current technology includes the extraction from image data of certain pertinent features of an image. In particular, image processors trace the significant edges of an image and hence build up a brief outline of the object field. From a set of outlines, it is sometimes possible to compute the most probable object associated with the set of edges and then proceed to determine object orientation in relation to the sensor (for example, a TV camera).

Image processes of this type are in use on industrial production lines and can handle limited sets of objects and orientations. Having deduced an object type and orientation, a production automaton can then proceed to institute object manipulation and so participate in the construction of a multicomponent device.

Multilevel image reduction is a technique in which the outline characteristics of an object field are deduced and stored, then a more detailed analysis can be performed using the first-level information as a basis for progressive refinement of the details. A number of such levels of detail can be used in successive correlation processes rather than attempt to search a vast information base for each minor detail. Such techniques are used in the computer world under such headings as 'Associative Data Base Systems'.

11.5 MULTILEVEL MEMORY SYSTEMS IN CORRELATION

In the world of analog devices, we have seen the delay line in a number of forms, the Miller-effect data store, CCDs and the more familiar surface magnetostatic and disc devices. In the digital world there are multiplane core stacks, RAM and ROM assemblies, CCD, disc, tape, and drum devices.

The computer uses multilevel memory systems – registers, ROM, RAM, core, disc, tape, and drum stores. The internal memory of many computers is of the split-level type consisting of a main memory and a fast cache memory. The principal reason is to achieve high-speed operation by repetitious use of the fast cache memory. However, to use this mode, data must be transfered from main memory to cache at the speed of main memory, making the decision to use cache dependent upon the type of work in progress.

Rather slow development has been evidenced for the long awaited 'content addressible memory' A noteworthy feature of the CAM is its ability to locate stored information according to a portion of the required set of information. Rather than the normal process of storage at a given location in memory, the CAM can determine the whereabouts of a set of information given a subset of it. This is a step along the way to an associative memory in which 'instant recall' of items associated with a given item is attained.

11.6 THE PULSATORY CORRELATOR

Given two cloud streams in adjacent matrices, it is conceivable that a third matrix could be coupled such that clouds were formed in the third matrix only when a high degree of correlation existed between the two source matrices. Various geometries are possible for this triad but we shall consider a general case in which a common interface is formed between the three matrices in some manner which can give rise to the correlation signal being set up.

For correlation to occur, there must be time coincidence of the two 'signals' being compared. It is most likely that the purpose of such a correlator would be to compare some incoming signal with a pre-stored signal sequence, that is, a signal preprocessor matrix and a memory matrix using some cyclic storage technique.

For any form of time coincidence to be possible, the signal preprocessor matrix should be a short-term memory capable of correlation with a cyclic memory in either time or space access mode.

For correlation in time, the incoming signal must be present at the correlation interface until the comparable memory cloud cycles past. If the memory is to contain a reasonable amount of information, then the time taken for a full memory cycle to occur could be excessive, requiring an exceptionally long signal matrix. The situation could be improved if the signal were routed into a short-term cyclic memory operating at a different cycle time from the main memory loop.

The alternative may be to arrange that the signal matrix couples over an extended length of the memory matrix. With such an arrangement, we can see that the nature of the stored information must suit that of the information emerging from the face of the preprocessor matrix. This in turn, determines much of the requirements for the insertion of memory patterns into the main memory system.

Yet another possibility is to use a spiral form of memory such that information can be accessed for correlation at different rates according to the radius of a particular section of the spiral. A number of conical forms of memory system suggest themselves.

Figure 11.1 takes the concepts a stage further. The matrix *a* indicates the memory image injector which involves a fairly powerful and synchronous signal injector. Matrix *b* is a normal signal propagative matrix which is coupled rather more loosely than is *a* to the ring memory. The coupling between the *b* and *c* matrices is very loose, amounting to virtual decoupling via the transverse memory matrix. However, when a cloud sweeps past the *b* and *c* interface region, if a suitable signal pattern is present at the *b* interface then the local activity level of the memory ring is increased. As a result, signal injection can occur into the *c* matrix. This could result in propagation through *c* and into the remainder of the system.

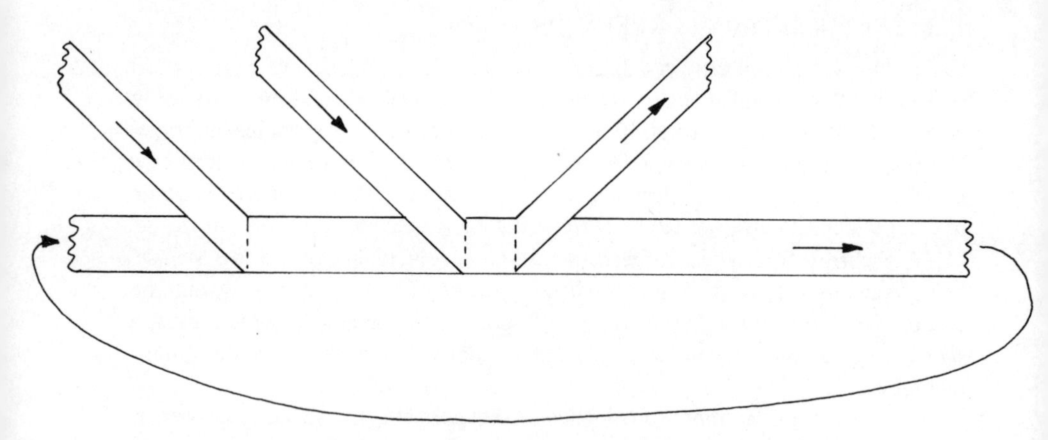

Fig. 11.1 – A Cyclic Memory with Transverse Correlator.

This provides a form of correlation amounting almost to memory retrieval. Because the signal launched into matrix *c* is due to the coincidence of patterns in the memory and the input matrix (*b*), the signal launched into matrix *c* will bear a close resemblance to that in matrix *b*. This amounts to saying that when an incoming signal correlates with the memory pattern, this signal will be passed to the output.

The fidelity of such a gated transmission path is limited. Where correlation is less than 'high', the signal amplitude at output will be of low value. Propagation through the matrix c is dependent upon signal intensity as has been indicated in Chapter 8. We thus have a correlator which can accept some forms of degradation in both signal path and memory images. The output is either a signal similar to the input (and stored) image or nothing.

In general, aliassing problems will be present to a certain extent but, by the nature of the pulsatory action (clouds with seeded trails), aliasses will normally be expected to decay within the output (c) matrix.

Correlation by computer is commonly a very slow process. Each byte of information in the signal must be compared sequentially with each byte (or other information unit) of the reference stream until a match is found. Analog correlators are commonly restricted in their areas of application. The searching of a large store of analog information can be an exceptionally long process.

Given the characteristics of the pulsor matrix, it would appear that there is likely to be a long way to go in the design of a large scale correlator. However, a number of approaches are possible.

Commence with a multilevel memory consisting of a very large 'backing' store, a sequence of 'intermediate' storage rings and the appropriate inter-device coupling matrices. Arrange that an incoming sequence can be transferred to a short-term memory ring. Also arrange that information segments from the backing store memory can be transferred to intermediate short-term rings and we see the beginnings of a system in which on-line information can be routed to a sequence of devices wherein correlation can be performed.

There are various possible outcomes from a correlation procedure. One is that a signal indicating correlation can be derived. Such a result can indicate the degree of confidence attributable to the process. Another possible outcome is that, from a partial message, the correlator can effectively retrieve from the backing store, a version of the 'whole message'. This is the effect of the CAM system.

Given a required outcome, we can design a system which will derive the wanted effect. To date, only the most elementary forms of correlation have been simulated. The matrices and signal used have been small and simple in nature and the observations have been of the graphical representations of activity in the simulator described in Chapter 12.

12

Design techniques – computer models

12.1 THE NEED FOR A SIMULATOR

A major problem in the design of pulsatory matrices is the fact that we have few mathematical processes to use in the derivation of fundamental design parameters. Much of this work must be of an empirical nature. Early work in this field was conducted using a few elemental circuits. Later, some hundreds of elements were fabricated using discrete components. Experiments with a number of forms of interconnection fields served to confirm that methods other than manual wiring would be required.

Much of the difficulty lies in the quantities of elements and the complexity of interconnection fields. We would look for a practical device containing many thousands of elements for rudimentary tests. For operational use, many millions of elements can be expected. By 1970, matrices holding perhaps a thousand million elements were under consideration. Microfabrication is the obvious technique for production – but what is it that we should fabricate? How do we determine the exceptionally complex interconnection field? The concept of using a computer for modelling was considered from the earliest days in this work, but the stumbling block was the effective size of the computer. Computers were used extensively in examination of many of the regular matrices introduced in the earlier chapters but computers hold information in discrete 'packets' – bits, nibbles, bytes, words, pages, files, and so on. No computer existed (nor yet exists) to hold and process the quantity of information involved in a pulsor matrix.

12.2 THE MEANING OF GENERATE IN RELATION TO A COMPUTER

It has long been recognised that sets of 'test data' could be 'generated' against a statistical specification by use of a computer. The compilers used in translation from a high-level programming language to 'machine code' include the concept of 'generating' the machine code sequences. During the NASA APOLLO project, it was concluded that programs to test the hardware of computers cannot be 'written'; such programs could however, be 'generated' in the computer being tested. There are then, certain classes of entity which can be 'generated' against some form of specification. Such work often relates particularly to the preparation of vast quantities of information.

An interesting feature of many of the stochastic information generators is the small size of the generator program. This permits the generation in a given computer, of larger data sets than would be the case if the generative program were large.

In 1975 came the realization that a generative technique for certain types of model could be used dynamically. Rather than attempting to generate an entire model, if just a small region of the model were generated and exercised at a time, a given computer could handle a much larger model of a system. It is a matter of using time as an alternative to space. Use of space-time dualities is an exceedingly powerful technique in computing.

12.3 SIMULATOR = MODEL + DATA

A computer simulation may be described as a model (in algorithmic or heuristic form), supplied with data and operated in time. Simulations may be time-compressive, real-time or in expanded time. The geologist can compress geological time into a short time-scale for the investigation of geological models. The air-traffic controller can use a real-time model of an airway:radar situation for training purposes and a cosmologist can attempt to model his 'first few microseconds' in extended time for the development of Big Bang theories.

The concept of Model Generation must be fitted into this simulation pattern. Firstly if the purpose of model generation is to minimize on-line storage requirements then we replace space (storage) by time (the modelling algorithm). Using this technique, just that portion of the model (a few pulsor elements and their interconnections) will be generated and executed. The algorithm then calls for another set of pulsor elements from the generator.

In a more recent development, a vast DESCRIPTION of the model is generated during the initialization of the model. During program execution (simulation), instead of wasting time by generating a portion of the model from the parametric data, the model description is used. The description is far smaller than the model in information content and hence storage. This provides some of the advantages of model generation without the serious disadvantage of extensive runtime.

12.4 SEQUENCING BY COMPUTER

The three principal techniques for computer operation are:

(a) Systematic progression through a data set or model
(b) Heuristic or stochastic sequencing
(c) Recursive sequencing.

In the first of these, we would progress step-by-step through the state-space matrix (array) of the model using strictly regular algorithms. To implement a stochastic search of our matrix we would select an element 'at random' and check its state, taking any action necessary. For the recursive technique, we would allow each active element to lead us to the next according to the state of affected elements following a 'trigger'.

Each of these techniques has been tried in practice and the current form of the simulator uses a mix of methods (a) and (c). A systematic search is instigated and recursive behaviour is entered as required by the state of each element examined. To produce computer programs of this order of complexity, either direct machine coding (or 'assembler coding') or a modern high-level language such as ALGOL is required.

The computer to be used must be 'permanently available' for 'single-user' tasks – and good on-line graphical facilities are essential. These requirements rule-out all mainframe computers, most minis and also most micros. The combination chosen after exhaustive surveys of the possibilities is the RML 380Z computer with high-resolution graphics.

12.5 PROGRAM CONTROL STRUCTURE

The control system for the simulator is based on the need for two operating phases – the construction of the model description and the simulator exercizer. For the first, control is 'linear' in that the user is given prompts regarding parameters required by the program. If the reply to any request is the value zero, a default value is used by the machine. For example a 'standard' matrix of 75 × 35 LSEs is set up if a 'length' of zero is requested. This enables a 'standard' experimental base to be set up rapidly. For the second phase, a more complex control system is required.

The principal process in the second phase is simply a matrix scan with recursive calls of the 'trigger' procedure as required. So the control system of the program must provide access to a number of parameter-setting algorithms and to the matrix processor. On entry to phase two, the user is presented with a menu of possible selections. By this, the user may adjust such parameters as the 'pulse power', and trigger level. Also included in the menu is the possibility to preseed the entire LSE matrix to any desired level, thus providing a base for many classes of experiment.

12.6 THE STATE-SPACE MATRIX

Figure 12.1 illustrates the basis of the pulsor matrix. The element at position v,w is assumed to have been examined and is in a state such that it should trigger. Other pulsors within the small rectangle are liable to be affected by the action of pulsor (v,w). The generated description of the model will disclose how

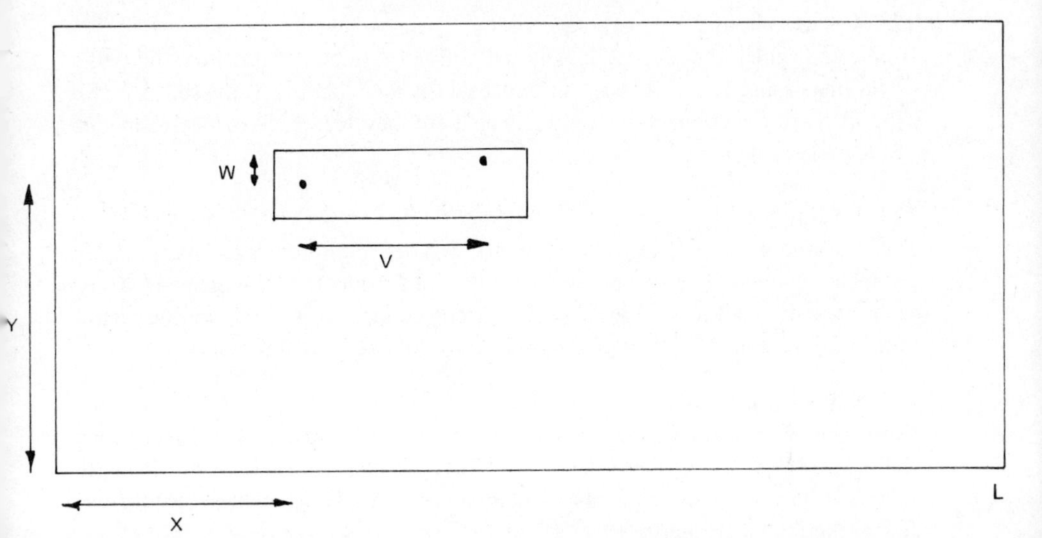

Fig. 12.1 – Active LSE in a Simulated Matrix.

many pulsors are to receive an impulse and their individual relative locations (x,y). Only the information relating to the set of elements to be stimulated by element v,w will be accessed and used by the modelling algorithm at a given time.

In use, the graphical display traces the interconnection pathways used by the simulator as impulses are produced. This enables the user to observe the behaviour of the simulator.

12.7 MODELS OF MATRIX PARAMETERS

12.7.1 Size

This is specified by the user in terms of length and width of the required matrix. The result is a planar matrix having 'width' LSEs in each of 'length' rows (a row simulating a plane in the physical matrix).

12.7.2 Bypass Length

The user provides a figure such as 20 or 30 which implies that the resulting interconnections could extend to that many planes (rows) forward. The distribution of such lengths is however, non-uniform. An approximation to a Gaussian distribution is achieved by summing a sequence of random numbers such that the maximum extent is the figure given, the mean is half that and the peak is at the position of the mean.

12.7.3 Transverse Spread

This is modelled in a similar manner to bypass length but the mean is zero as is the peak. The extent may be set to ±2, 3, 4 etc.

12.7.4 Backcoupling

If selected, then the bypass length has unity subtracted from it so that the origin, mean and full extent are shifted back by one 'plane'. As a result, a small proportion of the connections from any element may feed back to local elements of the preceeding row.

12.7.5 Complexity

The user supplies a figure, say 20 or 40, and the computer provides an equiprobable figure within the range 0 to the given value for each element. Thus each element is allotted a number of output connections which will be distributed as determined by the bypass, transverse and backcoupling values.

12.7.6 Reversed Elements

From the percentage figure given by the user, the computer will ensure that that proportion of elements has its 'X' coordinates negated. Such elements then will have interconnections with elements of preceeding 'planes' save for the few backcoupled elements.

12.7.7 Defective Elements

The resulting matrix descriptions provided by the computer will enable the study of varying degrees of complexity in interconnection fields and their results on the performance of matrices.

However, to accord with the Positive Principle, the simulator produces many defective LSEs, i.e. LSEs which cannot be triggered. The resulting simulations operate well despite these inbuilt defects.

12.8 MODELS OF LSEs

12.8.1 Characteristics

The 'triggering level', percentage of inhibitory operations and pulse amplitude may be respecified during simulation so that one can assess the behaviour of different LSEs under given circumstances. The summatory leakage parameter has normally been treated as a constant, it can readily be reset as a user-controllable variable.

Provision has been made in the simulator to integrate the activity within a number of regions of a matrix such that the sensitivity of LSEs may be controlled according to the local activity. Thus the behaviour of a matrix supplied from a poorly regulated power supply or supplied via resistive lines can be assessed. Such parameters may well prove of use in practical matrix processors.

The algorithms used for the activation and triggering of LSEs correspond closely with the performance of actual electronic circuits. Model LSE accuracy is of the order of 1% in relation to the real circuits. Because the prime design criterion is that no parameter need be of high accuracy and there must be provision for defective components, it was felt that the 1% modelling accuracy was satisfactory. To achieve significantly higher accuracies involves vastly increased computer time during simulation. The present simulator using all the techniques described here operates some thousand times faster than the original of 1975, contains many more features and provides a greatly improved form of graphical display.

12.8.2 Models of sensor elements

These are simply normal elements treated slightly differently during simulation. A sensor element is one situated on the input plane (or row) and which is excited to triggering level by the application of an input signal segment. Thus whereas an ordinary LSE has incremental pulse inputs which must be summed (with decremental action in time) to achieve trigger state, an input or sensor element is simply activated in a single step to raise it to trigger level.

The nature of the work undertaken to date has decreed that all inputs are go-nogo. A strong signal causes a number of adjacent sensor elements to be fully activated (pulsing frequently) whilst a weak signal activates separated elements. The reason for this is the time constraint in using a conventional computer – to activate a sensor element at variable rate would (in fact does) take a very long time.

12.8.3 Models of outputs from pulsor matrices

Output elements are treated slightly differently from normal elements. In the models and simulations, we visualize an LSE as having a number of OUTPUTS with interconnections to other LSEs. A trigger pulse is fed as an activating pulse to the inputs of other LSEs where pulse summation must occur.

The output element is viewed as an LSE which RECEIVES inputs from a number of elements near the output plane (or row). The state of excitation of the output elements is displayed graphically in terms of brightness levels at points on the display matrix. Thus, in use of the simulator one can observe the space-time integrand of activity near the output plane (or row). This is in contrast with the manner of representation of other activities in the matrix.

12.9 GRAPHICAL PRESENTATIONS BY THE SIMULATOR

In the initial stages, the displays are simply verbal to enable the user to design and control the experiment. However, during execution of an activation, high-resolution raster graphical displays are used. An area of the CRT screen is mapped to correspond with the matrix of LSEs which has been produced. Within this area, a bright spot is presented at the position of an LSE which reaches trigger level. Each LSE which is excited by the triggering element is denoted by (a) a feint line joining the positions of the two elements and (b) a medium-bright spot indicating the position of the excited element.

The resulting display provides a dynamic view of the activities within the matrix in terms of those elements which are excited and those which trigger, showing the active interconnections between them.

As time proceeds, there is a tendency for the screen to fill with a history of activity. At intervals, the screen is cleared so that only very recent and current activity is displayed.

From the displays, one can observe the nature and distribution of interconnections (when the screen is relatively clear) and the areas of the matrix affected by activity. The signal (when present) is displayed as a number of bright points at the left of the screen and the integrated output as a relatively diffuse patch at the right.

A reverse-connected element is displayed as a flashing bright spot.

12.10 PULSENET COMPUTER PROGRAM

The program on pp. 152–155 is just one of the many versions of Pulsenet which have been produced and exercised to date. This version, written as ALGOL60 (for RML 380Z computer), includes many of the features discussed in this chapter. The sequence of crt photographs of Fig. 12.2 was obtained using the program.

As with any specialized computer equipment, a number of non-standard terms appear in the listing. Reference to the 'ioc' and 'emt' procedures relate to the particular manner of accessing certain of the I/O devices and their associated control programs. The procedures 'resolution', 'point', and 'line' relate to the graphical display of dots and lines. The symbol '*N' in the 'text' procedures denote 'commence a new line'. Devices '1' and '7' in the 'write' and 'read' procedures denote the crt screen and the keyboard respectively. The particular version of ALGOL60 includes the 'BYTE ARRAY' – a variant on the normal 'INTEGER ARRAY' facility. This feature, together with the natural recursive nature of ALGOL60 was one of the deciding factors in selection of this particular computer.

12.11 EXERCIZING THE SIMULATOR

The sequence of photographs in Fig. 12.2 illustrates a typical start-up sequence for a simple experiment. The initialization phase takes a couple of minutes wherein the computer calculates large quantities of numbers representing the interconnections for a representative matrix. These numbers form the description used by the computer to generate the model dynamically during phase two.

The first of the photographs displays the activity in an unseeded matrix upon application of a 'signal'. Next is the state of the matrix after the signal has been present for a while. The third picture shows the behaviour following removal of the signal. After a short period with the signal shuttered, the matrix can appear as in the fourth frame. Finally is an indication of a typical response to signal reinstatement.

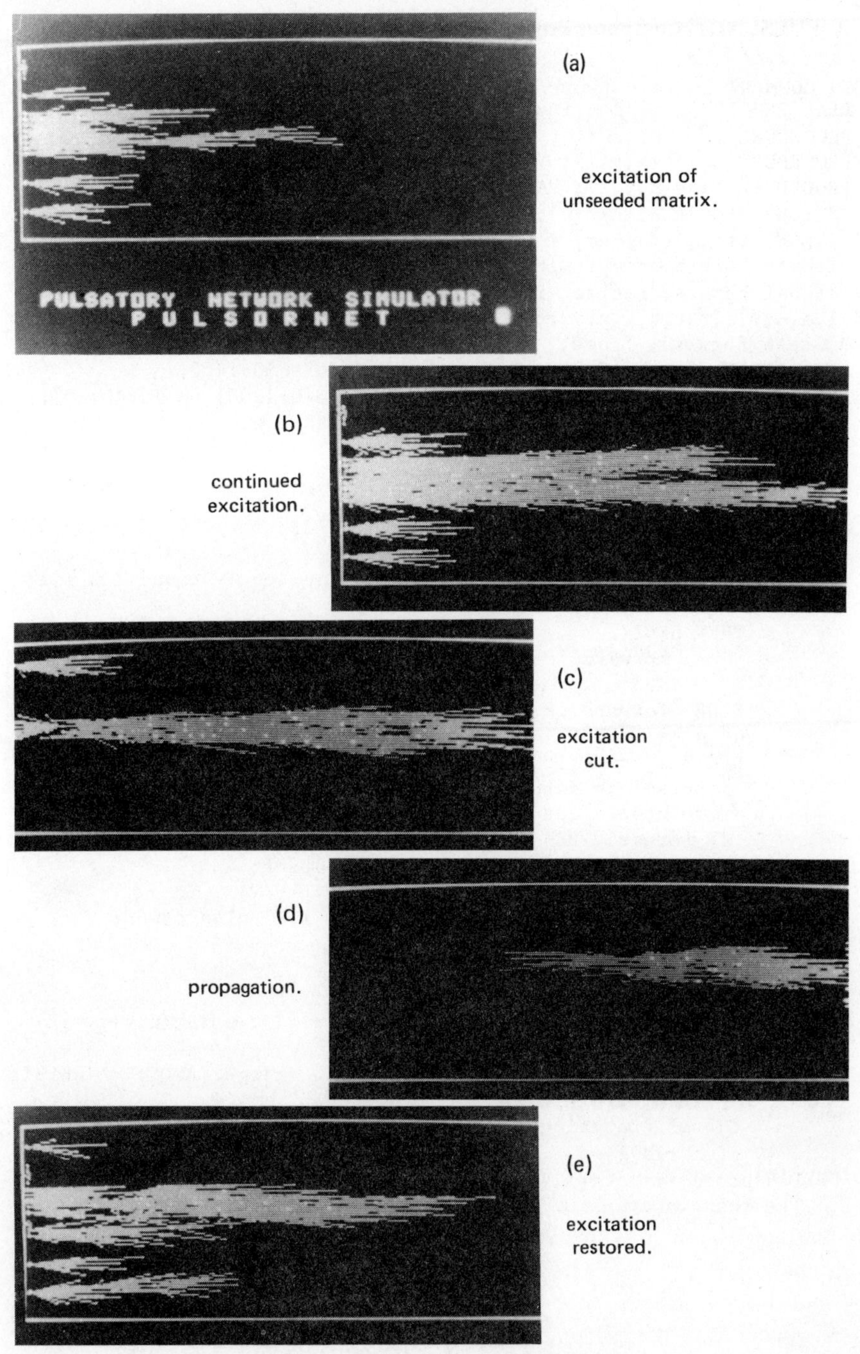

Fig. 12.2 – Snapshots from the Simulator.

A 'PULSENET' Computer Program – PULSER

```
BEGIN COMMENT pulse network simulator - PULSER;
   REAL PROCEDURE random;ioc(46);
   PROCEDURE resolution(i,j); VALUE i,j; INTEGER i,j; ioc(63);
   PROCEDURE point(x,y,i); VALUE x,y,i; INTEGER x,y,i; ioc(65);
   PROCEDURE line(x,y,i); VALUE x,y,i; INTEGER x,y,i; ioc(66);
   INTEGER PROCEDURE emt(n); VALUE n; INTEGER n; ioc(25);
   INTEGER trig, trigger, inc, dec, inpulse, bypass, pulse power,
    length, width, complexity, scan, quantity, connexion,
    signal bursts, cycles, inhibit, feedback, reversed,
    l,m,n,r,s,t,u,v,w,x,y,z;
   BOOLEAN forward, lines;
   BYTE ARRAY pulsor[-5:200,-5:30],model inpulse[0:1200],
    model connector[0:1200], model transverse[0:1200],sp int[0:10];
   PROCEDURE fire(v,w,z); VALUE v,w,z; INTEGER v,w,z;
       BEGIN a trigger:
          z := 0; pulsor[v,w] := 0;
          inpulse := inpulse MASK 1023;
          n := w%6; sp int[n] := sp int[n] + 1;
          connexion := connexion MASK 1023;
          quantity := modelc[connexion]; connexion := connexion + 1;
          IF quantity > 128
           THEN BEGIN
                 forward := FALSE; quantity := quantity - 128
                END
           ELSE  forward := TRUE;
          FOR n := 1 STEP 1 UNTIL quantity DO
           BEGIN a connexion:
            inpulse  := inpulse + 1;
            t := modeli[inpulse]; u := modelt[inpulse];
            IF forward THEN x := v + t - feedback
                       ELSE x := v - t + feedback;
            IF x < -4 THEN  x := 1; y := w + u - 2;
            t := pulsor[x,y]; pulsor[x,y] := t + pulse power;
            point(v,w+w+w,3);
            IF lines THEN
             BEGIN
              point(v+1,w+w+w,1); line(x,y+y+y,1); point(x,y+y+y,2)
             END;
            IF forward AND x < length AND t > trigger AND t < inhibit
              THEN fire(x,y,z);
           END of connexions;
       END of a trigger;
    SWITCH process:=settri,setinh,setsen, setinc,setdec,
          setstim,continue,setrun,terminate;
```

```
initialization:
   y := emt(13); y := emt(14); resolution(0,2); lines := TRUE;
   text(1,"*N*N*N*N*N*N*N*N*N*N*N*N        P U L S O R N E T*N");
   text(1,"*N*N*N*N*N Pulsatory   Network   Simulator");
   text(1,"*N*N   First build a model network:-");
   text(1,"*N*NGive matrix size ,'length','width' ");
   length := read(7); IF length < 20 OR length > 180 THEN
            BEGIN length := 75; width := 30 END ELSE
   width := read(7); IF width < 20 OR width > 25 THEN width := 25;
   text(1,"*NLength is "); write(1,length);
   text(1,"     Width is "); write(1,width);
   bypass := 12;
   text(1,"*NBypass length 10-30 "); write(1,bypass);
   bypass := read(7); IF bypass < 5 OR bypass > 30 THEN bypass := 12;
   write(1,bypass); complexity := 20;
   text(1,"*NComplexity 10-to-30 "); write(1,complexity);
   complexity := read(7); IF complexity < 5 OR complexity > 40
                       THEN complexity := 20;
   write(1,complexity);
   text(1,"*NLocal feedback '0' or '1' "); feedback := read(7);
     feedback := IF feedback = 0 THEN 0 ELSE 1;
   text(1,"*NReversed elements % "); reversed := read(7)*12;
   text(1,"*N  Your Model is under construction");
   FOR n := 0 STEP 1 UNTIL 10 DO sp int[n] := 0;
  FOR x := -5 STEP 1 UNTIL 200 DO
   FOR y := -5 STEP 1 UNTIL 30 DO pulsor[x,y] := 0;
  FOR n := 0 STEP 1 UNTIL 1200 DO
   BEGIN
     modelc[n] := (random+random)*complexity;
     modeli[n] := (random+random)*bypass;
     modelt[n] := (random+random)*2;
   END;
   FOR n := 1 STEP 1 UNTIL reversed DO
    BEGIN pulselement reversals:
     y := random*1200; modelc[y] := modelc[y] + 128;
    END of pulselement reversals;
    l := 1; m := 0; n := 0; pulse power := 20; inc := 0;
    dec := 0; inhibit := 120; trig := 100; trigger := 100;
    inpulse := 1; connexion := 1; signal bursts := 4; cycles := 1;


select process:
   y := emt(14); resolution(0,2);
   text(1,"*N   P  U  L  S  O  R  N  E  T ");
   text(1,"*N*N NETWORK SPECIFICATION");
   text(1,"*NNetwork Length ");write(1,length);
   text(1,"    Width ");write(1,width);
   text(1,"*NBypass length "); write(1,bypass);
   text(1,"*NComplexity (inputs-per-pulsor) "); write(1,complexity);
   text(1,"*NLateral spread +-2");
   text(1,"*NLocal feedback ");
    IF feedback = 0 THEN text(1,"NOT "); text(1,"used");
   text(1,"*NPercent reversed pulsors "); write(1, reversed%12);
   text(1,"*N*NCONTROLLABLE PULSOR PARAMETERS");
```

```
  text(1,"*N1. Triggering level  (ref.100%) "); write(1,trigger);
  text(1,"*N2. Inhibition level  (ref.100%) "); write(1,inhibit);
  text(1,"*N3. Pulse power             %    "); write(1,pulse power);
  text(1,"*N4. Stabilizing increment %      "); write(1,inc);
  text(1,"*N5. Stabilising decrement %      "); write(1,dec);
  text(1,"*N*NCONTROL OF EXPERIMENT");
  text(1,"*N6. Signal Bursts (n in 10) "); write(1,signal bursts);
  text(1,"*N7. Continue for one run");
  text(1,"*N8. Cycles per run "); write(1,cycles);
  text(1,"*N9. Terminate and Restart");
  text(1,"*N*N   SELECT ITEM 1-to-9 ");
 n := read(7); y := emt(13); resolution(0,2);
 GOTO process[n];


continue: text(1,"*N*N");
          text(1,"*N  PULSATORY  NETWORK  SIMULATOR");
          text(1,"*N        P U L S O R N E T     ");
 FOR m := 1 STEP 1 UNTIL cycles DO
  BEGIN
   FOR scan := 1 STEP 1 UNTIL 10 DO
    BEGIN a matrix scan:
     FOR n := 0 STEP 1 UNTIL 10 DO sp int[n] := 0;
     y := emt(13); resolution(0,2); trigger := trig;
     x := length+24; y := width*3 + 10;
     point(0,0,2); line(x,0,2); line(x,y,2); line(0,y,2); line(0,0,2);
     FOR n := 1 STEP 1 UNTIL width DO
      FOR v := length - 2 STEP 1 UNTIL length + 2 DO
       BEGIN output integration: l := pulsor[v,n];
        point(v+20, n+n+n, l%25); pulsor[v,n] := l*9%10;
       END of output integration;
     IF scan <= signal bursts THEN
        FOR n := 5 STEP 5 UNTIL width, 15 STEP 1 UNTIL 22 DO
              pulsor[1,n] := inhibit - 1;
    FOR v := 1 STEP 1 UNTIL 15, 17 STEP 1 UNTIL length DO
      BEGIN a lateral scan:
        FOR w := 1 STEP 1 UNTIL width DO
         BEGIN to check for a trigger:
          z := pulsor[v,w];
          IF z > inhibit THEN point(v, w+w+w, 0) ELSE
          IF z > trigger THEN  fire(v,w,z)
            ELSE pulsor[v,w] := z*19%20;
         END of a trigger;
         n := w%6; IF sp int[n] > 20 THEN
             BEGIN integration:
              sp int[n] := sp int[n] - 10;
                      trigger := trigger + inc;
             END ELSE trigger := trigger - dec;
        trigger := (trig + trigger)%2;
      END of a lateral scan;
    END of a matrix scan;
  END of one cycle;
 GOTO select;
```

```
settri:   text(1,"Trigger level 80-to-150 "); write(1,trig);
   trig := read(7);
    IF trig < 80 OR trig > 150 THEN trig := 100;
   trigger := trig;
   write(1,trigger);
   GOTO select;
setinh:   text(1,"Inhibit level % "); write(1,inhibit);
   inhibit := read(7);
    IF inhibit < 100 OR inhibit > 254 THEN inhibit := 200;
   write(1,inhibit);
   GOTO select;
setsen:   text(1,"Sensitivity 15-to-30 "); write(1,pulse power);
   pulse power := read(7);
    IF pulse power < 5 OR pulse power > 30 THEN pulse power := 22;
   write(1,pulse power);
   GOTO select;
setinc:   text(1,"Stability increment "); write(1,inc);
   inc := read(7);
    IF inc > 50 OR inc < 0 THEN inc := 0;
   write(1,inc);
   GOTO select;
setdec:   text(1,"Stability decrement "); write(1,dec);
   dec := read(7);
    IF dec > 10 OR dec < 0 THEN dec := 0;
   write(1,dec);
   GOTO select;
setstim:  text(1,"Signal bursts n-in-10 ");
   signal bursts := read(7);
   signal bursts := IF signal bursts < 10 THEN signal bursts ELSE 5;
   text(1,"Flush the Matrix '1' or '0'"); n := read(7);
   IF n = 1 THEN FOR x := 1 STEP 1 UNTIL length DO
                 FOR y := 1 STEP 1 UNTIL width DO
                     pulsor[x,y] := pulsor[x,y] + pulse power;
   GOTO select;
setrun:   text(1,"*NRun how many cycles ? ");
   cycles := read(7);
   text(1,"*NLines on display 1/0 ");
   n := read(7);
   lines := IF n = 1 THEN TRUE ELSE FALSE;
   GOTO select;
terminate: GOTO initialization;
END
FINISH
```

As yet, the simulator is used in a watch-the-screen mode. This is not so boring as it may sound. Following a period of interactive use, it is quite normal to 'take five' (for coffee) whilst things develop to an interesting stage in the machine.

The computer unfortunately is but a compromise. It is certainly not the right machine for this work – as yet it is the only machine we have which can permit active study of pulsor matrices.

12.12 USE OF PULSENET

12.12.1 Checking Theories

One of the early difficulties in the potential design of pulsatory matrices was that it appeared that signals would either diverge and flood or converge and die. It seemed that the transmission of a facsimile of an input with any details present was a most unlikely event. Much time was wasted with conventional mathematical research and with disappointing simulations. The problems surrounded the stabilization of matrix gain at unity when the gain itself was signal-level dependent.

However, by judicious juggling with sets of parameters and in particular, the inhibition ratio, the new simulator quickly indicated the route to follow for the transmission of self-stabilizing images.

This indicates the use of the simulator in directing research into the design and the behaviour of pulsor matrices.

12.12.2 Deducing Design Parameters

As with the application of the simulator in theoretical work, so the simulator can be used at least to provide parametric values to be used in the design of various types of pulsatory matrix. Signal situations may be set up, matrices may be seeded and then the LSE parameters adjusted such that operation of the simulator discloses for example, optimization of the levels for interacting parameters.

For a given 'design', the effects of electronic deterioration may be assessed to gain confidence in design criterior for real matrices having required performance. In fact, PULSENET and its family of variants may be used in the normal manner of simulation – as a base for theory-checking, parametric design and of course, for the demonstration of the nature of pulsatory matrices.

12.13 FURTHER DEVELOPMENT OF PULSENET

Using computers we are very close to the limits of what can be achieved. Following a total redesign of the PULSENET program using assembler coding we may improve by a factor of ten. Then it may be possible to incorporate videographics with the pulsor matrix memory in a totally redesigned computer – this could take some years of development work and the speed improvement is only expected to be by a factor of two or three. Faster microcomputers are not 'just around the corner' and the use of multiple computers would appear to be of little advantage – especially as this would represent a major computer design task which could take some years.

The current state of the art then is that we have optimized virtually every aspect of the simulation save the major program redesign and can hence look forward to a speed gain of around 10 over the current ALGOL program system. One major advantage of ALGOL is that it does include the recursive properties required by the simulator, the code is relatively efficient and the RML ALGOL is fast in comparison with other systems for this class of work. Another advantage of ALGOL over the assembler code systems is the speed with which one can prepare extensively revised versions of the program for special experiments. Commonly, a program will be developed from the master text in a few minutes, a range of experiments will be conducted and within hours, the new program has served its purpose and come to the end of its usefulness. It is possible that the new PERQ computer with its inbuilt PASCAL and 1000-line raster graphics may be applicable to PULSENET work – however, its 1981 price is prohibitive.

13

Fabrication techniques

13.1 CONSTRUCTIONAL OVERVIEW

There is likely to be a variety of matrix forms used in the construction of pulsor devices and systems. These will range from relatively simple arrangements on a single silicon slice to large multislice assemblages. The memory and processing power associated with a pulsor device will range from that of a single-channel signal processor to that of a national communications network.

13.2 SILICON STRUCTURES

Commencing with the smaller-scale devices, we would expect to see relatively simple rectangular silicon slabs with sensor devices at one edge, a relatively powerful processor offering unidirectional propagation and outputs available at other edges. Figure 13.1 illustrates the scheme, with a second possibility as in Fig. 13.2. Such devices, based upon the simulation studies of Chapter 12 may be used singly or in groups. Chemical interconnection fields are the most likely and will permit cooling and purification by pumped fluids.

At the next level, one would expect to see the widespread use of complete silicon slices. One of the exciting fields in design of such devices is that of the compound-function slice.

Imagine a silicon disc such as is the start-point of todays microelectronic technology. Construct a sensor region, for example, a patch of photosensors at the centre, use radial propagation leading through an anular memory region, as in Fig. 13.3.

The resulting structure could be designed to provide active and dynamic pattern recognition.

The system could have programmable learning capability and would represent an extremely adaptive control system.

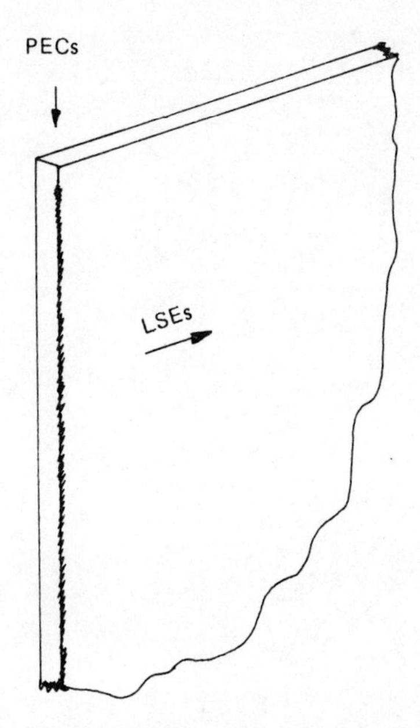

Fig. 13.1 – Row of Photosensors on a Plane of LSEs.

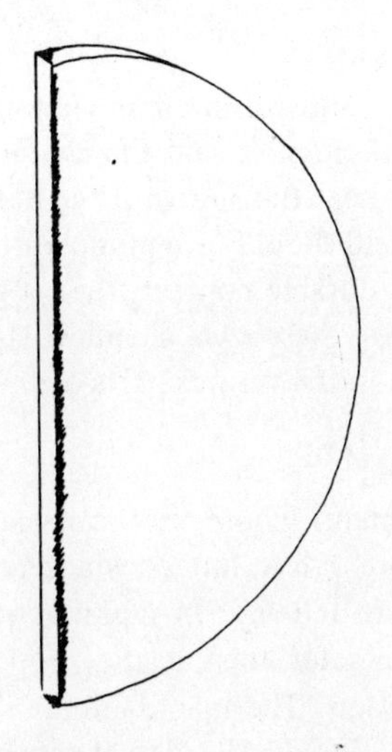

Fig. 13.2 – Half-slice photosensing LSE plane.

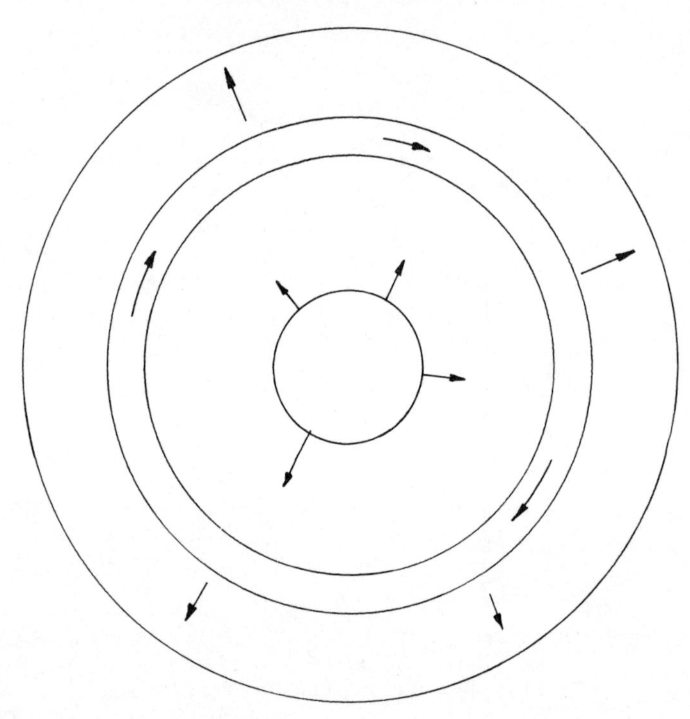

Fig. 13.3 – Radial disc with Central Sensor and Anuluar Memory.

13.2.1 Interconnection Fields

The most promising of the forms of interconnection fields is that of the chemical conductor. Many types are possible and the characteristics are varied. To combine the requirements for heat transfer with the need to prevent undesirable cell formation, the chemical field should be a pumped fluid.

To ensure stable and durable contact, then a grease form would be appropriate. The combination of these two chemical states could be used to confer required characteristics upon the various parts of the system.

13.2.2 Manufacturing Yield

To ensure a failproof system, whose performance is unaffected by elemental failures, we need to ensure a low but constant rate of failure. Pulsor systems with defective elements are reliable. In manufacture then, defective slices are acceptable. A design parameter then is the proportion of defective elements per slice on initial inspection. The manufacturer should aim towards a reasonable defect rate per slice. Should the defect rate per slice be too high, it may be assumed that performance of the slice will deteriorate at an unacceptable rate in service. As always, if the inspection acceptance level is set too high, then the yield in production falls too low.

A further consideration however, is that, if a pulsor matrix has too high a proportion of good elements, then its activity level may become unacceptably high in relation to that of systems of considerable service age. We would prefer then, that possibly as many as 25 or 30 percent of produced elements are defective on initial inspection to ensure that the matrices have the required performance characteristics.

13.3 STATIC CHEMICAL INTERCONNECTION FIELDS

The two most obvious forms which could be taken by stationary chemical interconnections are:

(a) Static liquids.
(b) Greasy chemicals.

In either case, they would form resistive electrical paths between cell connections on a slice or electrochemical cellular connections between points of a slice.

In the one case, an electrical field set up by a triggering pulsor would spread out through an electrolyte to deliver elemental charges to adjacent pulsors. With this system, the field strength due to a pulsor would diminish rapidly with distance. Furthermore, the quantity of charge delivered to other pulsors would have a variable value. Nearby pulsors would be highly excited whilst remote elements would be less affected. Such effects can readily be studied using the simulator of Chapter 12.

In the case of polar electrolytes, various forms of cellular activity can be set up either at the output or at the input points on the face of a slice.

Interface chemicals can be designed to provide for either stable or variable performance according to such parameters as temperature or electrical activity. Designs may be for such purposes as activity stabilization or automatic sensitivity control. Again, the simulator makes provision for studies of the effects of such parametric control. Whilst it does not provide detailed design criterior for development of chemical interfaces, the simulator does provide the means for studies of the effects of providing or not providing such control.

13.4 DYNAMICAL CHEMICAL INTERCONNECTION FIELDS

With this form of interconnection we open an entire new world of design possibilities. Our slices will generate heat – why not use a fluid to remove that heat from the pulsatory region out to an external radiator or converter? Meantime, the fluid could also behave as a chemical interconnection system.

The possibilities opened up by this arrangement include the 'cleanup' of the electrochemical interfaces, prevention of the developments of activity hotspots and a further degree of independence of the system performance upon individual system components.

For the stabilization of electrochemical cellular activity based upon the pulsatory actions, timescales have to be relatively long. The constant removal and replacement of the conductive fluid can be used as a further performance-stabilization agent. Now there is little time for the setup of low-rate actions. There is no question of the setup of 'preferred paths' as may well be the case when using stationary fluids. In fact, we cannot tell which elements are connected to which at successive points in time. An independence has been achieved in the map of element-to-element 'wiring'. Not only do we no longer know or care which actual interconnections are used – we now find them altering with time.

Heat extracted from the slices can be converted to electrical form and used to bolster the power supply in much the manner by which the classical TV raster power supplies are operated. By this approach, smaller power supplies may be used than would otherwise be the case. If the power demand of the slices is dependent upon activity then the circulatory interconnection fluid could provide further stabilization and economy of power supply.

The power gridding to convey the working potentials across a slice can be used to constrain the chemical interconnections to the required areas of the slice. By such methods as this, the propagation pathways may be constrained into 'bundles' and the various active regions may be defined.

13.5 SLICE ORGANIZATION

13.5.1 Vectored Radial Propagation

A possible line of development is the radial disc of Fig. 13.4, take alternate discs with opposite radial flow vectors. The first could be the sensor disc having a central sensor patch and outward radiality. Follow this by an adjacent disc with inward radiality. Coupling from the first to the second such disc could be by fluid connections at the peripheries.

The centre of the second disc (and subsequent discs) could be punctured with a reasonably large hole. From the second to the third disc, coupling would be at the disc central region. Thus alternate discs have different radial flow directions, coupling alternates between peripheral and central regions with the possibility of remote bypassing over whole discs, that is, from the peripheries of discs 1-2 to 3-4 etc. and from the centres of 2-3 to 4-5 etc. This form of extended bypass coupled with annular memory on each disc would have very great attractions for processor designers.

13.5.2 Orthogonal Propagation

Another very promising geometry for silicon slice design is the punctured disc. Given minute laser-drilled holes over the surface of a set of discs, we could visualize a possible signal flow path through the disc rather than basically across their surfaces. Now we have design bases rather closer to the rectangular crystaline structures considered earlier in the book.

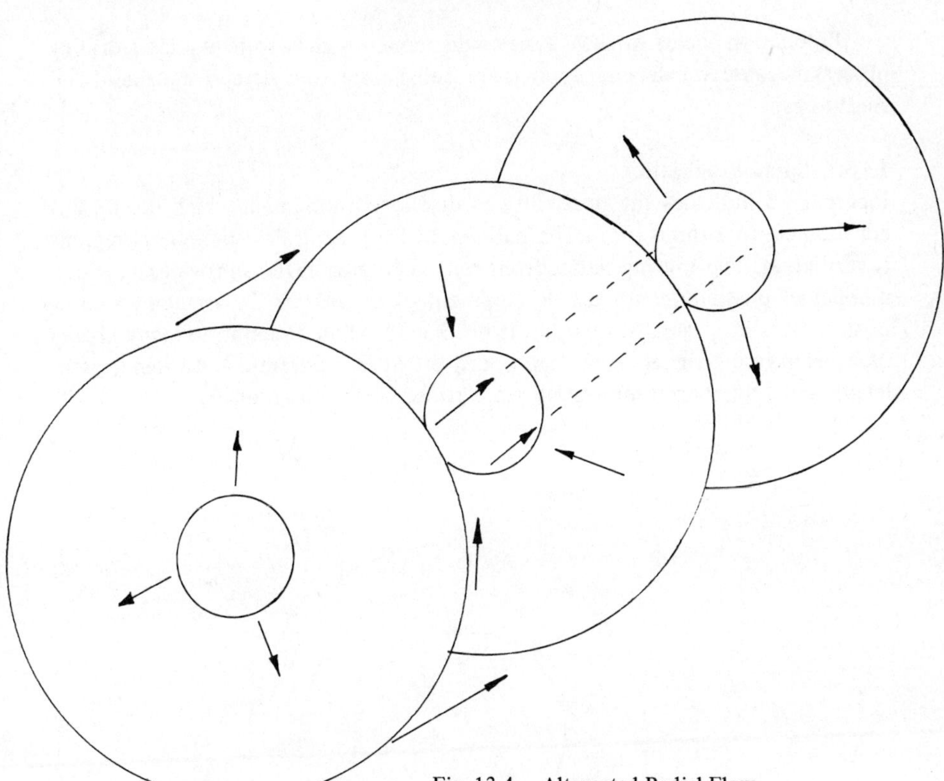

Fig. 13.4 – Alternated Radial Flow.

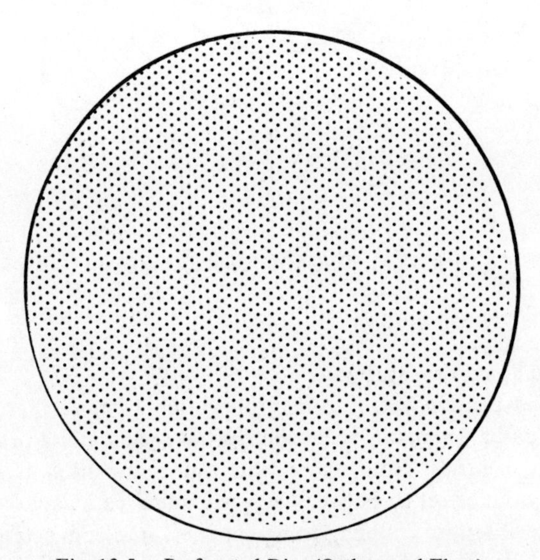

Fig. 13.5 – Perforated Disc (Orthogonal Flow).

Production yields of such discs could remain high despite the laser drilling process because we depend upon there being numerous defective elements on fresh discs.

13.5.3 Spiral Propagation

Figure 13.6 indicates the possibility of disc constructions in which the flow is constrained to a spiral form. The pathlength from central to peripheral regions is very great. The varying radius from central to edge regions provides us with a number of possibilities for the development of correlators. By arranging a series of interfaces in a radial formation, it becomes possible to tap in to many phases of a propagating signal. Each 'tap point' provides a different local signal pathlength and a number of interesting correlation possibilities arise.

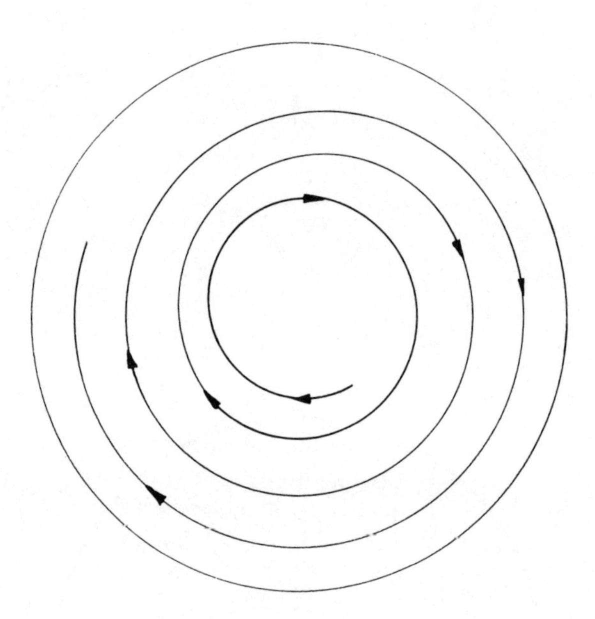

Fig. 13.6 – Disc with Spiral Propagation.

13.5.4 Multiple Spiral Propagation

Figure 13.7 takes the concept of spiral propagation a step further. Now we have multiple signal paths emanating from the central region and a new set of correlation possibilities arises. Both space and time correlation is now possible. Correlation can even be set up between adjacent spiral pathways.

Face-to-face interdisc coupling using chemical interconnection fields opens the way for a whole range of signal flow geometries.

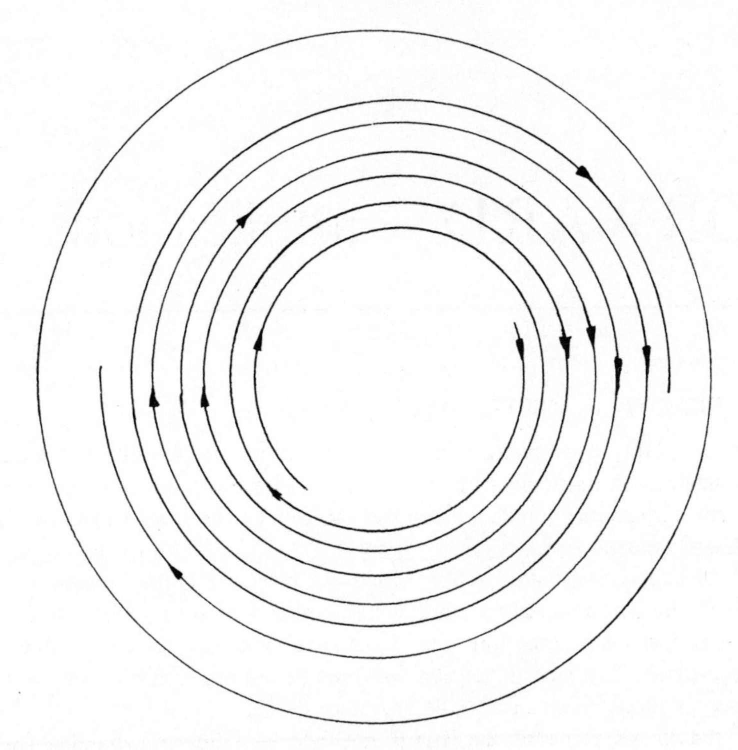

Fig. 13.7 – Double Spiral Propagation.

13.6 ELECTROCHEMICAL LSE

During 1970 a pilot study was conducted of the potential applications of chemical technology to pulsatory circuit design. A number of areas of application were examined, in particular, the possibility of using electrochemical cells to form the leaky summator section of pulsatory elements. Broadly, the conclusions reached indicated that such cells were simple to design and construct, were reliable in operation and could be made to exhibit most of the phenomena discussed here.

With the devices constructed, dependence on activity could be designed-in such that a cell could be highly sensitive until the onset of high-rate activity at which point the sensitivity could be encouraged to decrease obediently. Long-term tests were not undertaken at that time, so no precautions against atmospheric contamination were incorporated. In fact, checks made of those same devices a couple of years later disclosed the underlying lack of good design.

Another feature of the electrochemical leaky summators was that electrochemical time constants bear little relationship with the pulse times we aim for in microminiaturization. The tests were conducted using millisecond pulses.

14

FORWARD – applications

14.1 FIBROPTIC IMAGE TRANSFORMATIONS

A number of very convenient image transformations can be obtained by purely optical devices. In particular, a combination of a lens with its associated components and a fibroptic bundle can be used to reduce the spacial dimensionality of an image space.

The lens has a focussing action which can project a limited region of space normal to the lens axis onto a plane which is also normal to the lens axis. Thus, the lens acts as a selector within the object space and provides a projection onto an image plane. The time dimension can then be used to scan the external field, so passing reduced image data to the processor device.

If the image plane of the lens is the face of a fibroptic bundle, further spacial reduction can be achieved. The nominally circular receiving face of the fibroptic bundle can be routed through the bundle so as to be transformed into one of many other possible arrangements. Particularly, the output end of the bundle can be distorted as to produce a linear rather than planar arrangement. There are many possible planar disc to linear transformations and projections; some of these are particularly suited to certain of the silicon fabrication arrangements for pulsor matrices. Figure 14.1 indicates the arrangement.

Some of the possible spiral transforms are illustrated in Figs. 14.2–4. The first is a simple arrangement in which pixels from the centre of the planar face of the bundle are taken to one end of the linear output. A spiral trace from the centre of the receiving plane is mapped pixel by pixel to the output line. This provides one possible interface to an edge-input matrix. Again, the linear projection from the image plane may itself be wrapped around the edge of an edge-sensory disc. Here, the matrix must include LED outputs.

In the second diagram, Fig. 14.3, the spirolinear transform is taken a stage further by producing two pixel-interleaved lineal outputs. These can be applied to a pair of silicon devices. In the third illustration, Fig. 14.4, multiple lineal outputs are obtained from a single plane by appropriate redirectioning of the individual fibres of the bundle. This provides a number of correlation possibilities according to the design of the matrices used.

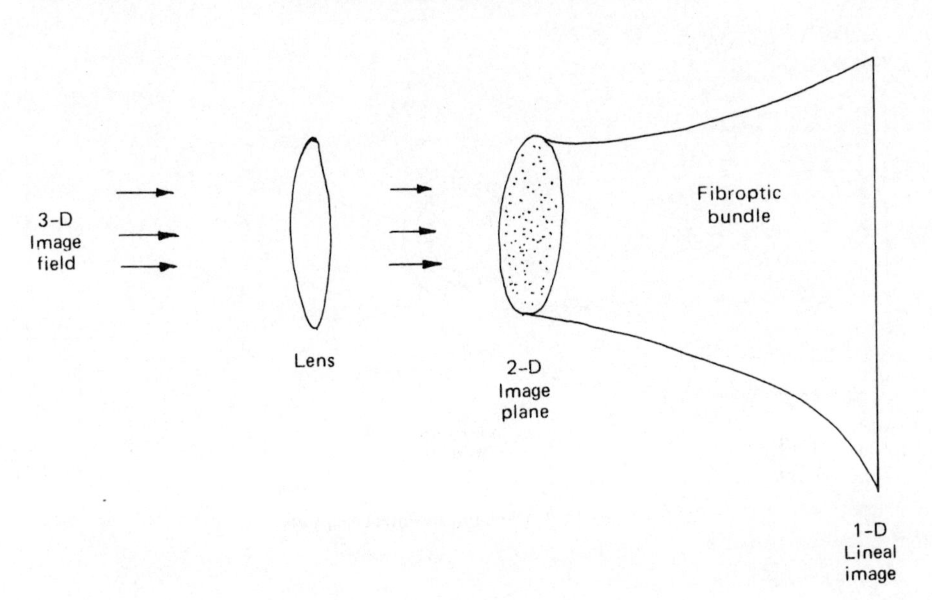

Fig. 14. 1 – Optical Dimension Reduction.

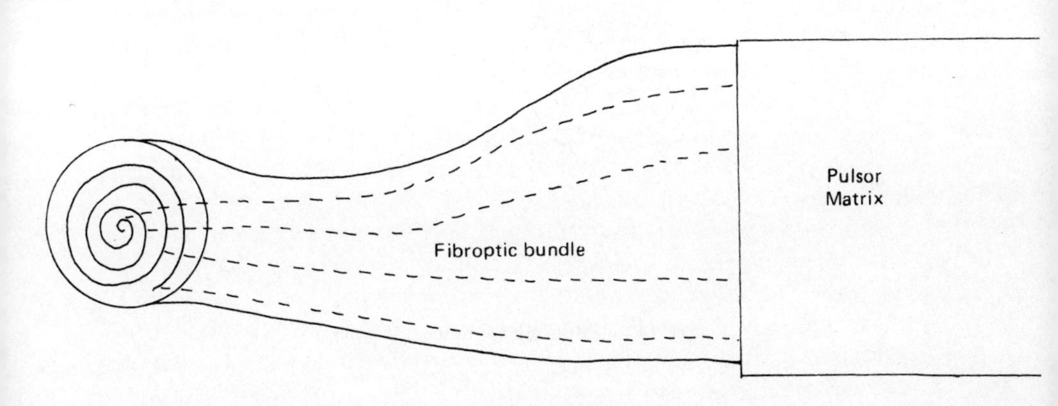

Fig. 14. 2 – Sample Spiro-lineal Mapping by Fibroptic Bundle.

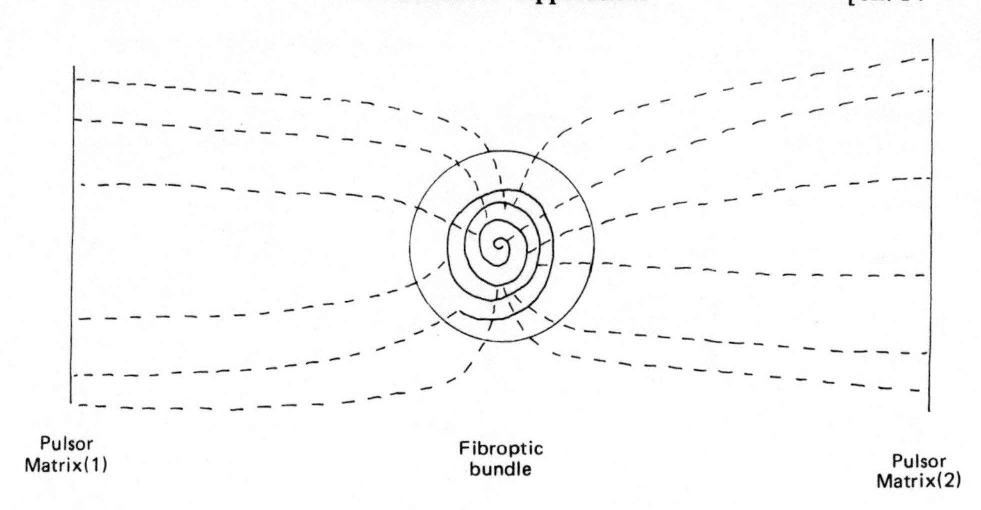

Fig. 14.3 – Dual Spiro-lineal Mapping.

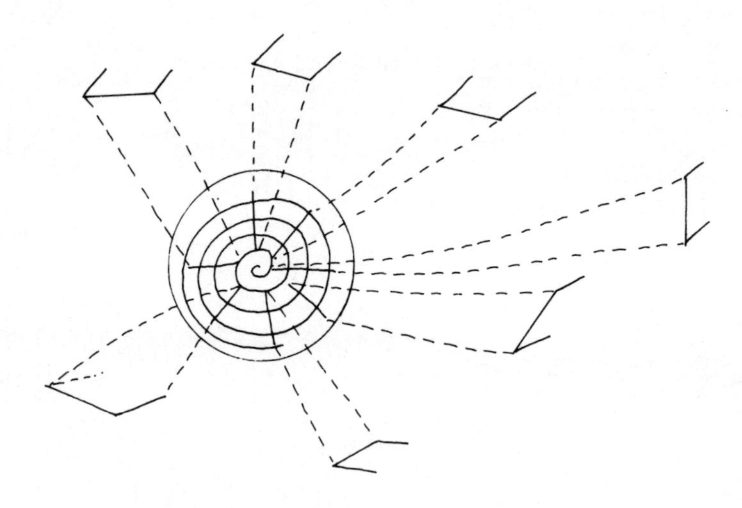

Fig. 14.4 – Multiple Spiro-lineal Mapping.

14.2 SERIAL DATA HANDLERS

In addition to the fundamental types of processor discussed in earlier chapters, we shall need to provide such devices as filters, resonators which can be operated in true parallel such as is the case with comb filters. Again, a spiral formation would appear to be well suited to such a task. Data could be introduced radially across a set of tracks having different pathlengths – and hence different response wavelengths. Outputs could be taken as a set of 'channels' each providing a response according to certain frequency components in the signal. The simple application of such a technique would result in filters responsive to sets of aliass frequencies – just as is the case with certain types of conventional digital filter and with many forms of resonator. However, by successive stages of filtering, such difficulties can normally be circumvented.

NOTE – reference has sometimes been made to 'frequency' and sometimes to 'wavelength' throughout the book. It is conventional to work in terms of frequency or 'normalized frequency' in relation to the processes of Fast Fourier Transforms. However, when dealing with filters and resonators, 'normalized wavelength' is often a far simpler concept to deal with. Frequently, the equations and manipulations necessary, become greatly simplified when transformed into wavelength terms. The use of the two terms then, has not been indiscriminate but appropriate in context.

14.3 PULSOR CONTROL SYSTEMS

A significant advantage offered by the pulsor system over conventional control techniques is the inherent autostabilization and self-optimization potential of the pulsor. Thus a new phase of control system development can be opened up. There is the possibility of linked multichannel control system stabilization such as is required by the automatic control of space vehicles. Current technology permits only the most elementary forms of inter-channel control communication.

Pulsor servo controllers can obviously be made relatively immune to impulsive noise and have the integration characteristics required for the minimization of random noise in both temporal and spacial domains. The multilevel memory systems of pulsor technology will permit the continuation of 'follow' during the temporaty absence of control signals. The concept of 'programming' changes to one of 'learning'.

14.4 AUTOMATA – INDUSTRIAL ROBOTICS

Industrial robots of the early 1980s give some considerable cause for concern in regard to reliability. In practice, some types of robot display a failure rate, and demand attention at rates which are far too high for economic comfort. This combined with the very high research, development and installation costs does make one enquire afresh whether it is not better to employ people in many production situations after all.

The incorporation of the 'TV eye' into the conventional computer-controlled robot brings many new problems, not the least of which is the increased complexity of the robotic installation. Whilst the Utility of the 'visualizing robot' exceeds that of the simple 'teach by guidance' system, the certainty of operation of the image-processor controlled robot may often be less than is desirable.

All too frequently, the introduction of advanced automation devices does not actually reduce the number of employees – it simply raises the technical background required of employees. This can lead to some of those sociological horrors predicted long ago by the science-fiction writers.

14.5 THE LEARNING AUTOMATON

One special feature of the pulsor system is its capacity to acquire and retain information regarding its environment. Given also, the function of correlation and hence recall, one has the potential for an automaton with inbuilt capability for spacial image processing. The speed of processing is basically far in excess of that provided by modern computers, and it offers the great advantage of being able to handle a complete image. This is due to its inherent processing capability.

One can therefore foresee an automaton which could 'read' a set of constructional drawings and use the information in the control of its associated assembly equipment. The concept of 'vision' is not unnatural with the pulsor system – so our industrial robot of the future can be expected to have full optical (even multi-optical) facilities for the exploration of its environment.

Using a number of compound sensor devices, the automaton could construct a sufficient base of 'world knowledge' to enable strategic decisions to be made in the course of normal operation.

14.6 PULSORS IN COMMUNICATIONS

For single-channel work such as audio and other relatively low-rate information, the pulsor matrix would appear to have little to offer over the conventional technologies. However, a number of problems remain in such fields and amongst these is the problem of message recognition. We still have no real mechanism for full translation of a message stream from one natural language to another.

The pulsor matrix may well be applied to such problems as full language translation. With the technology of the 1980s we still have no real way to overcome the problems of inter-language communication because of the limitations of both theory and practical devices. The storage capacity of the pulsor matrix can be extended well beyond that feasible with digital systems and the speed of processing by the pulsor matrix exceeds that of any conceivable digital system by orders of magnitude.

The softening of signals within the pulsor matrix makes possible the retrieval of fuzzy information, the high processing speed enables both syntax and context checks to be made in message decoding. The vast storage capability of the new matrices will enable true on-the-fly inter-language conversion.

14.7 SELF-REPAIR?

No attempt has been made to introduce this concept in the main body of the text. Rather, the emphasis has been on the development of devices which do not need repair. Not by use of redundancy in the conventional sense of the term – but by the development of an alternative technology. There are Analog and Digital systems – now we examine the possibility of the Pulsor system whose aim is to achieve reliability (and ultra-processing) in a novel manner.

To introduce self-repair is a topic which may have to await the arrival of full-blooded electrochemical cellular structures which may come nearer to the neurological world than does the projected pulsor system.

There has been some evidence of self-repair in silicon structures – but of an unwanted kind. Certain of the old 'fuse-link PROMS' could (sadly) regrow some of the conductive channels deliberately destroyed during programming. No, they wouldn't – would they? Could it be that Silicon Valley may use research into such 'unfailures' in attempts to produce a self-repairing device.

14.8 INTELLIGENT CONTROLLERS?

Well, forgive me if I sound a little less than enthusiastic on this point. We do not actually know what we mean by 'intelligent'. The many many 'intelligent computer' systems that have appeared over the past few years display a remarkable lack of what we think of as 'intelligence' in the animal kingdom. Do we even know what we mean by the relative intelligence of animals?

We should beware against the misuse of the term 'intelligence' amongst 'computer people'. There is a subject frequently referred to as 'Artificial Intelligence', whose startpoint is a definition which certainly does not describe the attribute which we understand as 'intelligence'. It describes a system which is remarkably close to a modern associative data base and bears some resemblance to what has come to be called an 'expert system'. For example, on an encounter with a medical man we find that the doctor will derive a number of parametric values by examination and relate these to knowledge built up over many years to derive either precise or 'most probable' diagnoses. A remarkably similar result can be obtained from the use of a medical automaton and an extensive associative data base of knowledge derived from the true medical experts. To imply that the automaton is 'intelligent' seems to be stretching fantasy to the limits.

Artificial intelligence then, is a misnomer of the first kind and the subject of computing abounds with such fallacies. We must beware of the use of natural-language terms which have been carried over into other disciplines and given new, restricted, technical 'meanings'. Think of 'intelligence' as being related to "that brainpower left over when the organism has been serviced". The warm-bloods have need of a great deal more neurological support than do the cold-bloods. The smile requires an incredible amount of muscle control – supplied neurologically. Speech may be an actual handicap – it requires considerable neurological activity. How does a bird compare with mankind on this view of intelligence? Actually, we do not show up very well. Could there be something more than we currently assume behind the "smile on the face of the Crocodile"?

Please do not confuse human failings and neurological deformities with the sort of behaviour we expect from pulsor matrices. We can choose and control the characteristic behaviour of these new automata. There will be no danger of ravages by the 'intelligent robot'. Even dear I. Asimov will not need to apply his 'three laws of robotics' to our machines. Pulsor matrices may be complex, operating on statistical principles, adaptable and so on, but we do not need to discuss relative intelligence levels in relation to them. At an official Faculty meeting in Oxford, the decision was taken unanimously – abandon the term Artificial Intelligence and substitute 'Advanced Automation'.

Index

A

activities, 23, 150
adaptive, 158
AGC, 107, 110
algorithm, 147
algorithms, 146
aliasing, 109, 110, 143
amplifier, 28, 69, 119
amplify, 32
analog, 13, 28, 29, 32, 33, 34, 130, 171
AND 58, 58, 59, 71
AND-END, 37, 43, 44, 54, 58, 60, 61
APOLLO, 20, 29, 30, 30, 33
assembler, 156
associative, 127, 129, 140, 141, 171
auto-corrective, 30
automation, 170
autostabilization, 169
autotest, 30, 32
availability, 13, 20, 23, 34

B

backcoupling, 94, 96, 99, 101, 102, 112, 119, 122, 148
benchmarking, 127
Boolean, 55
Buffery, 55
bundle, 90, 91, 107, 126, 166, 130, 131
burst, 109, 111, 132
bypass, 89, 94, 96, 96, 107, 115, 121, 147, 148
bypassing, 99, 162

C

capacity, 130
chemical, 160
cloud, 105, 106, 107, 107, 109, 117, 118, 119, 121, 124, 128, 132, 135, 141
clouds, 110, 111, 123, 129, 131
combinatorial, 55, 57, 59
complex, 25, 27, 29, 32, 32, 45, 48, 50
complexity, 13, 15, 17, 19, 20, 23, 24, 33, 34, 89, 102, 105, 112, 119, 123, 124, 144, 148
component, 27, 28, 29, 30, 36, 50, 54
components, 24, 25, 26, 28, 43
computer, 19, 29, 30, 32, 33, 36, 49, 101, 103, 104, 105, 110, 113, 125, 126, 127, 141
computers, 13, 17, 23, 27, 139
computing, 36, 145
constants, 56
control, 146, 169
converge, 156
convergent, 52, 83, 87, 88, 89
cooling, 158
correlation, 132, 139, 140, 141, 142, 143, 170, 171
correlator, 131, 140
correlators, 164
coupling, 162
cycling, 29, 34

D

decay, 102
defect, 28, 28, 30
defective, 70, 103, 112, 124, 148, 149, 160
degradation, 27, 27, 28, 124
De Morgan, 59
density, 129
description, 145, 146
deterioration, 26, 27, 156
Diagnostic, 23
differential, 108
differentiate, 32, 107
differentiation, 110

diffusion, 128
digital, 13, 28, 32, 33, 36, 50, 127, 130, 171
dimension, 166
discrimination, 112, 119, 123
display, 150
distribution, 147
diverge, 156
divergence, 102
divergent, 83, 87, 89
dualities, 145
dynamic, 150, 158
dynamically, 150

E

electric, 124, 125
electrochemical, 80, 104, 161, 162, 165
element, 28, 37, 40, 43, 50, 52, 54, 59, 61, 62, 64, 74, 82, 102
enhancement, 93, 94, 101, 102, 112, 123, 125
exercizing, 153, 155
experiment, 150

F

fabrication, 33, 50, 91
facsimile, 156
failproof, 62, 160
failure, 13, 25, 26, 27, 29, 36, 50, 54, 93, 94, 95
failures, 28, 28, 29, 54, 102, 103, 104, 160
false, 56, 57
feedback, 41, 103, 103, 112, 115
fibroptic, 104, 126, 166
field, 124, 125
filters, 168, 169
flood, 106, 109, 118, 156
flushing, 122, 132
focussing, 166
frequency, 169

G

gain, 59, 62, 63, 64, 64, 69, 102, 103, 106, 107, 109, 112, 116, 118, 119, 121, 135, 156
interrupt, 32
Gaussian, 87, 147
generate, 144, 150
generator, 145
glimpses, 121
glitches, 32, 33
graphical, 149, 150
gridding, 131
guides, 126

H

hardware, 29
heat, 160, 161, 162
hybrid, 33
hypercube, 46

I

image, 166
inhibit, 57, 78
inhibition, 107, 108, 109, 111, 122, 123, 156
inhibitory, 103, 112, 119, 148
initialization, 150
integrate, 32, 148
integration, 110, 121, 135
intelligent, 171
interacting, 156
interconnection, 36, 54, 90, 93, 94, 94, 104, 111, 144, 147, 148, 160
interconnections, 26, 40, 41, 50, 52, 87, 88, 95, 96, 101, 102, 107, 113, 145, 150, 162
intermittent, 27, 28
intermodulation, 128, 129, 130
intraplane, 94, 96, 101, 116

J

jitter, 121

L

ladder, 134, 135, 136, 137
lateral, 99
learning, 158
length, 147
linear, 128
logic, 13, 13, 48, 50, 55, 57, 70, 71, 74
logical, 43, 56, 57, 58, 62, 82
lse, 63, 64, 69, 75, 76, 77, 78, 80, 90, 93, 105, 107, 110, 116, 117, 122, 125, 130, 131, 146, 147, 148, 149, 156, 165

M

magnetic, 124, 125
mainframes, 126
majority, 71
majority-logic, 61
manufacture, 160
marginal, 23
matrix, 26, 36, 50, 52, 53
memory, 28, 30, 45, 49, 93, 101, 102, 109, 124, 128, 129, 130, 131, 131, 132, 133, 135, 136, 137, 138, 141, 142, 143, 158, 162

MFLOPs, 126
microcomputer, 126
microfabrication, 26
MIOTs, 126
MIP, 126
model, 104, 112, 145
modelling, 101, 144, 147, 149
MTBF, 13, 20, 23, 24, 34, 35
MTTR, 13, 23, 24, 34
multilayered, 138
multislice, 158

N

NAND, 58, 71
network, 95, 96, 96, 101, 102, 103
Noise, 111
NOR, 71
normal, 111, 171
NOT, 57, 58, 58, 71

O

operators, 56, 57
optical, 125, 126, 166
optimized, 157
OR, 58, 59
OR-END, 60, 61
orthagonal, 162
orthogonal, 49

P

parametric, 145, 146, 156
particles, 125
penetration, 94, 99, 102, 103, 107, 119
performance, 23
pixels, 166
Poisson, 87, 96, 111
power, 162
probability, 48, 55, 57, 58, 60, 62, 64, 64, 69, 94, 107, 108
programs, 32, 36, 127
projection, 166
propagates, 119
propagation, 52, 86, 93, 103, 106, 107, 115, 116, 117, 121, 122, 123, 129, 158, 162
propagative, 89, 134
pulsatory, 50, 70, 80, 87
Pulsenet, 150
pulsensor, 50
pulsor, 50, 56, 74, 171
pulsorcube, 38, 41, 42, 43, 55, 60, 61
pulsorcubes, 37, 45
pulsorhypercube, 45, 46, 48
pulsormatrices, 52, 53

R

radial, 162
radiant, 124, 125
random, 13, 147
real-time, 145
recognition, 140, 158
recording, 130
recursive, 145, 146, 150, 157
redundancy, 30, 32, 36, 93
re-entrancy, 134
reflective, 134
reflexive, 128
reliability, 24, 25, 26, 27, 30, 30, 33, 33, 34, 36, 37, 50, 90, 95, 104, 116
repair, 29
representation, 149
resolution, 104, 121, 130, 132
resonate, 32
resonator, 169
resonators, 168
retrieval, 139, 142
reversal, 102, 119
reverse-connected, 150
reversed, 94, 96, 101, 113, 115, 116, 119, 122, 123, 148
RsDP, 23

S

safety, 30
scanning, 125
screening, 125
seeded, 119, 121, 156
seeding, 122, 128, 135
self-stabilizing, 156
sensitivity, 119, 121, 122, 148, 165
sensor, 149
sequencing, 145
sequential, 13
Shannon, 55
shielding, 125
shuttered, 119, 122, 150
shuttering, 109, 110, 121, 123, 125, 131
signal-processing, 15
simulate, 103
simulation, 110, 111, 112, 117, 133, 145, 156
simulator, 127, 146, 147, 153, 155
size, 147
slice, 113, 158, 161
slices, 162
software, 29
solid, 27
specification, 144, 145
spiral, 142, 164, 166, 168
spread, 99, 147
stability, 33, 53

stabilization, 103, 105, 106, 107, 109, 156, 162
statement, 171
state-space, 146, 146
stochastic, 145, 146
summator, 62, 63, 104
supercomputers, 126
superpulsorcube, 45
symbolism, 55
systematic, 146

T

tolerance, 70
trail, 106, 119, 123, 128, 129
transformation, 166
transients, 32, 32
transverse, 128, 132, 148
trigger, 102, 106, 132, 146, 149
triggering, 123, 148
true, 56, 57, 59

U

unreliability, 20
unseeded, 150
usefulness, 13, 17, 19, 20, 23, 24
utility, 20, 23, 24, 30, 33, 34, 35

V

validation, 29
variables, 56
vectors, 162
velocity, 122, 123

W

wave, 124
wavelength, 169

X

XOR, 57, 58, 60

Y

yield, 69, 70, 160
yields, 164